Small Coal and Smoke Rings

Small Coal and Smoke Rings

A Fireman on the Great Western

DEREK BROCK

John Murray

First published 1983
by John Murray (Publishers) Ltd
50 Albemarle Street, London W1X 4BD

Typeset by Inforum Ltd, Portsmouth
Printed and bound in Great Britain
by the Pitman Press, Bath

British Library Cataloguing in Publication Data
Brock, Derek
Small coal and smoke rings.
1. British Rail. *Western Region*
2. Locomotive firemen—England
I. Title
625.2′61′0924 TJ603.G7
ISBN 0–7195–4076–3

Contents

Illustrations

Between pages 70 and 71

Illustrations

Maps

SOURCES OF ILLUSTRATIONS

The author and publisher warmly thank all those who helped during the picture research for this book: in particular, for his continual encouragement and advice, Dr Stuart Owen-Jones of the Welsh Industrial & Maritime Museum, Cardiff. Sources are as follows: 1, the author; 2, 3, 4, Welsh Industrial & Maritime Museum; 5, 6, 11, 12, 18, 19, J. Pitt; 7, 9, 10, E. R. Mountford; 8, M. Macleod; 13, 15, 16, 17, R. O. Tuck; 14, G. Deighton; 20, M. Esau

Foreword

Many of today's railway enthusiasts profess an allegiance to railway companies which ceased well over thirty years ago at the time of nationalisation and so can have little appreciation of what conditions of work were really like under those regimes. Derek Brock's account is both of interest and of value through his description of daily conditions as experienced by one who did not enter in the belief that the initials GWR stood for God's Wonderful Railway or subscribe to the view that all the company's activities and actions were to be accepted, without question, as being for the benefit of greater mankind.

Small Coal and Smoke Rings makes clear in a most illuminating manner how, despite the strict regulations and the Rule Book which prescribe virtually all actions and decisions on a railway, and proscribe many others, the actual operations contained many departures from the set procedures. Some of these are credible, while others – for those who were privy only to platform-end observations of a railway – will be read with a sense of incredulity.

It is of interest to reflect that the interval between the author's joining Barry Shed in 1942 and the present day is over twice as long as that between 1942 and the grouping in 1922, when the Barry Railway (opened in 1889) formally lost its independent existence and became part of the Great Western Railway. Not surprisingly in 1942 there was still a strong Barry Railway flavour at the Shed, as the author makes clear, with many of the staff having actually worked for that company. Today, alas, virtually all evidence of it, from the massive viaducts and steel bridges to the enormous storage sidings at Cadoxton, has gone so that in the space of less than a century little remains of that once-great Welsh company.

Within the relatively short period of the author's railway career the book conveys vividly the flavour of the time, particularly with regard to the mineral trains, which were the life-blood of South Wales, and the local passenger trains which, if the engine was not in good condition, offered no respite for the fireman upon whom fell the full burden of persuading the engine to produce steam. The author also demonstrates how durable and tolerant the steam engine was and how strenuous was the life of a fireman, subject to the idiosyncrasies of the different drivers with whom he worked, by no means all of whom possessed an empathy with the engine which was being driven.

The book is pervaded throughout by a variety of those incidents which provide the spice of life and leave the reader with a wish that other railwaymen, former and present, may feel inclined to publish their own experiences and so enrich our knowledge of the human side of railways rather than the historical and mechanical aspects which have been extensively documented.

Stuart Owen-Jones
ASSISTANT KEEPER
WELSH INDUSTRIAL & MARITIME MUSEUM
CARDIFF

1 March 1983

Prologue

I placed the Sunday paper on the coffee table, then, with a quarter-turn of my swivel chair, glanced up at the ceiling in a clouded search of the past.

'Steam Nostalgia' was the heading of the article. It told of a wealthy company director who had become hooked on steam conservation. His addiction forced him to rise from his bed at the ungodly hour of six on a Sunday morning and drive his Rolls-Royce a journey of fifty miles to the engine shed of a steam preservation society. I smiled as I mentally re-read the article: 'He donned the overalls that transformed him from a property tycoon into a railway fireman. For one glorious day, he forsook the wheel and comfort of his Rolls for the shovel and spartan surround of the footplate.' My smile broadened at the last sentence: 'He paid for the holy privilege of being one of the highly selected chosen few.'

When I left the Great Western Railway as a mainline fireman some twenty years previous, with less than eight pounds a week in my pay packet, I would scarcely have classed myself as one of 'the highly selected chosen few'. I very much doubt that that company director would have changed places with me in those days, even on a Sunday, for steam railways were an established way of life, and the lot of a fireman was a rather mundane one. But the article had begun to arouse my imagination. I recalled, with a growing pride, my ten years on the iron road.

That afternoon I found myself wandering between the

rusted ghosts of Woodham's Yard at Barry Docks – by coincidence, my home town. The nostalgia was taking effect as I wandered through the lines of lifeless leviathans of the past. For the first time in more than two decades, I grasped the once-shining handrails of a locomotive – a Great Western 2800 class 2–8–0. As I subconsciously took hold of the rusted regulator, my nostalgia intensified, and the engine became alive. The triple odours of hot cylinder oil, acrid sulphur and coal dust filled my nostrils. A white-hot heat played on my legs from the firebox, and the hiss of live steam everywhere.

I thought of the company director, and my once-cynical intentions began to mellow. Although each weekend he made a glorious escape from his rat-race to the past, or nearly the past, I was more fortunate – I had done the real thing. I had earned my living on his cherished museum pieces.

I felt a compulsion to write of the railways – but what? Bookshop shelves sagged with countless journals on nearly every aspect of the days of steam. As I searched my mind for an idea, I chuckled to myself. I was a reluctant railwayman, as you will see, and I accepted the job in a light-hearted manner. Technically I was useless, for I never attended the Sunday morning Mutual Improvement Class, but as a matter of personal pride I took great pains in my steam-keeping, perhaps more in fear of the ultimate disgrace of 'sticking'.

A slight easterly wind began to increase causing the locomotive's loose sheet-metal boiler panels to rattle in the silent weed-covered sidings of Low Level. It was all so desolate, so ghostly, not even a preservation chap in sight.

I lowered myself to the ground with a mission in mind. The '28' was not my loco, for the fingers on one hand were more times than I had fired one, and the same applied to a 'King', 'Castle', 'Hall' or 'Manor'.

Prologue

My pace increased between the lines of towering locos in a futile search of 'Old Barrys', 'Rummies' and 'Taffies', but the scrappies' torches had already claimed them. 'Well, what next down memory lane?' said my spirit of nostalgia as I stopped at a GWR 5700 class pannier, one of the 67xx series fitted with a steam brake. 'Dozens of the little blighters at Barry Shed,' I thought, 'and the 5600 class 0–6–2 tankies, plenty of those too – everyone of them had a story along with the men who worked with them.' Once more I was on the ground in search of engine numbers that I knew. But there were no brass plates, just the whitewashed ownership numerals of the enthusiasts. I passed several graceful old ladies of LMS and SR vintage, but they were not part of my heritage.

I ended my search on a GWR 4200 class 2–8–0 tanky and felt sure there was enough material for a book. I looked into the near-empty bunker of crusted lifeless small coal which represented the bad days for a fireman. Then my eyes travelled along the boiler to the chimney-stack and the vision of a smoke ring. Like a shooting star they were a rare sight. What caused them I have no idea, but to me, and I suppose, to most firemen they were a fantasy symbol to wish on, just like a rainbow. It sounds a bit silly, but despite the hard work of the footplate, there was a lot of time for thinking, and to daydream over a smoke ring on a fine day when the loco was going well was no bad thing.

So there I had the title of the book: small coal for the bad days; smoke rings for the good days – and that's what made the job unique.

1 · The Reluctant Railwayman

'You're going on the railway, and that's that!' My mother paused to assess my reaction, then continued, 'What else is there for you? Telegram boy or the Maypole Stores. I'm ashamed of you after all that your Uncle Ned has done. Many a boy would give his right arm at the thought of becoming a driver like him.'

I gave a sigh with a shrug of defeat. All my pleas to avoid a career on the iron road had fallen on deaf ears. Mother could not afford the fifty pounds indenture money for an engineering apprenticeship, and my dream of being a sea-going engineer vanished swiftly into oblivion, as final arrangements were made for me to enter the service of the Great Western Railway Company as an engine cleaner.

My only ray of hope came from my sister, as the family planned my future during that wartime evening of 1942. 'There's no need to worry,' she said. 'You'll never pass the medical – you're too skinny.'

As the following morning's train carried me to my medical at Swindon, I had little doubt in my mind that the now despicable Uncle Ned had pulled strings. The GWR was indeed a family company, and despite the extravagant legends of the railway's constant search for supermen to match their thoroughbred locomotives, I suspect the lineage connection took first place.

Park House, Church Place, Swindon, the venue for the day-long medical, was a mansion-style building under the supreme command of Doctor Bennett, a thick-moustached man in his early fifties. In his black jacket and waistcoat with pinstripe trousers, he portrayed my vision of a Harley Street specialist. In fact, I soon learned from more knowledgeable candidates my first rather doubtful GWR legend.

Doctor Bennett had been procured from Harley Street by the Directors of the GWR to ensure that their footplate men were physically and mentally the cream of manhood! In later years, when I sometimes flagged under the strain of firing a bad loco, with poor coal, and too many nights at the local dance hall, I queried deeply the wisdom of his selection, and wished I had started my life as a mere human.

The medical was elementary, but extremely effective. Before confronting the great man came the various tasks to qualify. A battered cardboard box was placed before me. It contained strands of wool of every colour of the rainbow.

'Sort those out proper,' said Doctor Bennett's minion.

The lad next to me from Truro whispered in a superior manner, 'Colour-blind test.'

Next came a long wooden box with divisions, and a sliding lid. The minion smiled as he placed a jam jar of mixed coloured beads alongside the box. His smile intensified as he snapped the lid closed leaving just a neat little row of holes as the only entrance to the sections. With a full smile he spoke, as his eyes glistened with sadistic pleasure, 'There we are lad, sort 'em all out.'

The Truro boy whispered again, 'You got to remember where you put them, this is a colour-blind test and a head test, see.'

I survived the eye wall-chart with flying colours, but Doctor Bennett's assistant was not convinced.

The Company Rule Book was placed between my feet as I stood over it. 'Read that,' he said, with a slight air of dismay – the Truro lad and myself were going great guns, and the Doctor didn't like to end up with too many candidates; in fact we were to learn later that only three out of thirty-four candidates were to become engine cleaners.

'Is it correct', I said to the Truro boy, 'that failures from here have later joined the Police and the Guards?' He nodded his head in proud agreement and I soon learned that this was no legend.

I didn't, however, place my feet in wet sand as a flat foot detection as I had been informed – that was a yarn.

Doctor Bennett's room was on the first floor facing the wide flight of highly polished wooden stairs. The three finalists sat in nervous apprehension at the base. My prayers for failure had vanished during the elimination contest to get this far. Thirty-one of our number had failed, thirty-one mortals not good enough for God's Wonderful Railway, even in wartime.

The door suddenly exploded open to reveal the great man. 'Right, first lad run up, please, come on make it snappy, get your heart pumping, that's what it's for.' He pushed me into his room. 'Shirt and vest off, on the couch.' His stethoscope searched my body. 'Right, drop your trousers and cough, and again. Touch your toes.' His inspection torch scanned my ears, throat and eyes. His hands missed no part of my anatomy. He stook back in the middle of the room, his thumbs hooked into the armholes of his waistcoat as his fingers tapped each side of his chest. I could feel the penetration of his eyes as they mentally X-rayed me. 'I suppose you'll do for the Great Western, my lad, so settle up with Mr Ingram downstairs.' His index finger waggled near to my nose. 'And always remember this, lad. I can pass you physically fit for the Company, but I can't say that you're good enough for it.'

The Truro boy passed me on the stairs, as the Doctor barked his orders to the frightened candidates and I presented myself to Mr Ingram.

'You've passed then?' His smile was thoroughly forced.

'Wouldn't have come to me if you hadn't. Sign here, that's for your Rule Book, and here, that's to join the Great Western Mutual, and here's your registered number, that sticks with you as long as you're a servant of the Company.'

I waited outside for the Truro boy, who was too elated to speak as we walked back to Swindon station, and I wished only that I could conjure up a fraction of his enthusiasm as he interrupted our journey to telegram his family, who, he said, were awaiting his result with bated breaths.

Later, as the Cardiff-based 'Castle' headed me home, I did manage to create a weak fantasy that I was her fireman, but those days were as hazy as my tepid dream.

Uncle Ned arrived that evening, confident of my success, for he brought me two suits of overalls, a cap and a 'tommy tin'. The overalls and cap were to cause me later much discomfort, for they were Barry Railway vintage, and during my initiation period as a cleaner I suffered such remarks as 'You must have cleaned the Rocket.' I thought Uncle Ned had been a little premature with the 'tommy tin'. It was a handsome black enamelled model with brass fittings. The lid compartment contained a red flag, four detonators, a gauge glass, and rubber washers. He grasped my shoulders. 'Always keep your Rule Book in there as well, nipper.' I forced a smile of gratitude. I must have been the most reluctant servant of the GWR.

To a present-day enthusiast, Barry Running Shed would have passed as a portion of heaven, housing locomotives of four companies – the Rhymney Railway, the Barry Railway, the Taff Vale Railway, and the then current owners, the Great Western Railway. Pride of place, however, went to a temporary lodger, a giant 2–8–0 tender loco of the United States Army Transportation Corps. To this day I don't know why Barry was host to such a monster. For three months she berthed in No. 6 road, as cold as death, whilst

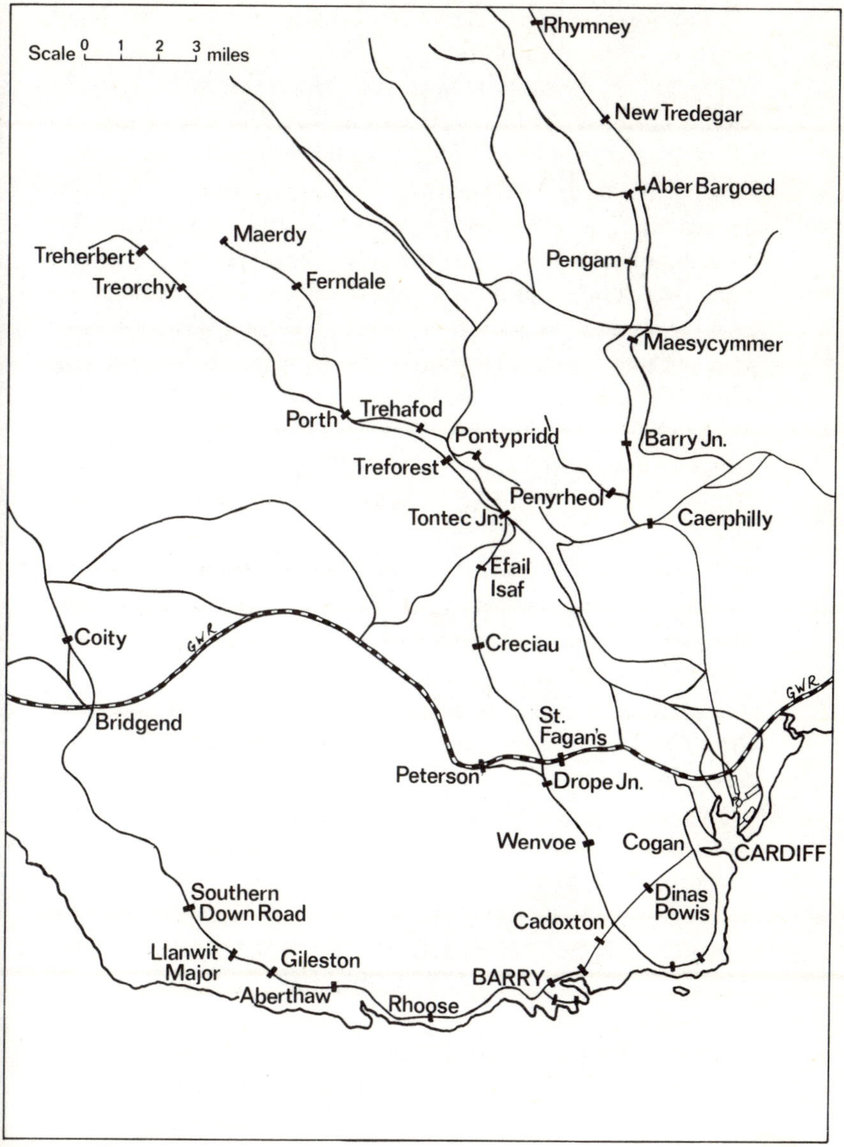

Main railway lines in Cardiff/Barry area

her side-rods rusted, and a cocoon of dust obscured all evidence of her nationality.

The ash path leading from the dock road to Barry loco shed must have a place in the mind of every engineman to use it. About nine feet wide, a few hundred yards long, it was built with a mixture of ash and clinker from the smokebox, firebox, and the ashpans of generations of locomotives. The crunch beneath your boots usually portrayed your physical condition when signing on or off duty – I have taken the path at a gallop when an early finish of an afternoon turn gave me the chance of an unexpected visit to the local dance. The path had so many associations as I trudged it every hour of the day, every day of the week, every day of the year. What engine will I have? I wonder what the coal is like? Hope my mate turns up, don't want a miserable old sod instead of him.

The shed at the end of the path had three main peaks of activity around 6.00 a.m., 2.00 p.m. and 10.00 p.m. At these times a constant queue of locos would come to life in each of the six roads as their crews got them ready. The ring of steel against steel, the hiss of steam, the bark of exhausts, the chorus of human voices, all created the atmosphere of a busy shed such as Barry.

At ten to six on a Monday morning, I trudged the path for the first time, and my boots seemed to tread the ashes with a nervous apprehension as I approached the roster room.

'Your number's 329,' snapped a little old man through the pigeon-hole. 'That's more important than your registered number.' He laughed. 'Can't draw your wages without knowing that.'

With some minutes to spare I walked over to the roster. Beautifully written in fine copperplate, with a spread the size of eight newspaper pages, it was locked securely in a light oak glass case. My eyes scanned the top link, the passengers, and down through the main line, main line spare, countless pilot links, light duty links, more spare links, to the very bottom of the ladder: Barry Sidings link.

As I walked past the drop pit towards the cleaners' cabin through an acrid haze of sulphur fumes, I paused to glance through the open door of the enginemen's cabin. A large copper boiler at the centre of the room held pride of place, and well-scrubbed wooden tables made the surround against the spotless concrete floor. The cleaners' cabin must have been the ultimate contrast. It was filthy and vandalised, a victim of the constant battles waged by high-spirited youths.

I have read and heard of the general code of conduct amongst public school boys, and in a working-class way, if there can be such a comparison, there was little difference in the behaviour of GWR engine cleaners. I soon learned that there was a strong code of honour, and no victim ever squealed to the foreman about his assailant. Seniority was taken very seriously, and every junior would automatically fag for his senior. The initiations for a newcomer were quite brutal in some cases. I recall how I was taken to a quiet part of the shed, to be stripped naked, then smeared with 'Blackjack', thick green cylinder oil, and finally garnished with fine sand. It took hours to cleanse myself with paraffin, and the whole episode had a marked negative effect on any slight enthusiasm that I may have possessed for my job.

The next day was my day of revenge. I took to work an old Great War hand-grenade, which had lain about the house for years and, to the best of my knowledge, was completely harmless. Still smarting on my tender parts from the abrasive action of the sand, I took my place in the mess room to await the arrival of Chippy German, the chargehand. I felt uncomfortable as the fifteen other lads connived with each other on my obvious fate. 'That can't be a Barry Railway cap!' said one. 'It's over twenty years since the Old Barry was in business, look how new it is.' I began to tremble as my heart raced, something awful was going to happen quite soon.

I jumped from my seat to reach into my conveniently placed 'tommy bag', and the act began. I had seen it many times before in films, now I was living the part. I stepped onto the table with the grenade in my hand. 'Right, you guys, one more move and we all go to hell!' My accent was wrong, decidedly Welsh, and vibrating with fear, but it had the desired affect. My fifteen potential assailants sat glued to their seats, wide-eyed with stark terror. I slowly pulled the pin. 'Now listen,' I said, as I slipped my thumb over the spring-loaded detonator pin, 'I'm fed up with you lot. Now clear the way, I'm going to let myself out, lock the door behind me, and if I'm still mad I'll lop this through the window.'

I felt a strong, yet nervous sense of power as I walked between them to make my exit, right into the arms of Chippy German. His eyes glanced casually at the menacing grenade, as he handed me the stores ticket for cleaning oil and cotton waste.

'Now stop playing bloody soldiers, and get some engines cleaned, you lot.'

While he presented stores tickets to the other senior hands, I realised that I, the miserable junior of the whole shed, held the same symbol of superiority. I offered the ticket to my senior hand, as Chippy left the cabin. He smiled pointedly at the grenade. 'Put the bloody pin back in that thing before we goes to the stores.' It wasn't long before I realised the power of my useless grenade, which I carried for many weeks to come.

In a predominantly tank engine shed such as Barry, the two juniors of a gang each cleaned a tank, half the bunker and cab, a set of wheels, a siderod and framing, while the two seniors had the simple task of half a boiler and smokebox, plus the motions. I am proud to record that I cleaned only the latter on my first engine — a 5600 class 0–6–2 tank.

After suffering two or three days of ritualistic initiation, involving countless journeys to the stores for such items as smokebox door keys and red oil for rear lamps, finally I was accepted.

My first social invitation was to a card school in the firebox of the 'Big Yank'. It was the ideal hiding place from Chippy German, the elderly chargehand, who constantly hounded us to clean engines to the standard desired. To the light of an oil ducklamp, we enjoyed many a game of rummy in the roomy grate of the 'Big Yank', but Chippy was relentless in his search, and the day of our discovery dawned eventually.

'Got you, you little blighters! Well, you can bloody well stay there till I fetch Mr Edwards, the foreman. It'll be up the path for you lot, an' serves you damn well right.'

With that, his beaming face vanished, as the doors closed and we became imprisoned, or so he'd thought, but the ingenuity of those cleaners was invincible. Firebars were lifted to allow our lithe bodies into the ashpan, where we slipped through the dampers and into the drop pit.

Ten minutes later, we arranged ourselves in the cab of an adjacent '56' class loco, as our would-be gaoler ambled ahead of the foreman through the half light of the smoke- and steam-filled shed. His voice shook with excitement, as he began to climb the steps of the American loco.

'Caught 'em red-handed, I did, Mr Edwards. Wondered where they'd been hiding for days, I have, then I finds the little devils in yer. I tell you, it's a regular Monte Carlo inside that firebox. See for yourself.' He opened the doors with a stage flourish. 'There we are, Mr Edwards.'

Our sides ached with suppressed laughter, as the foreman shone his torchlight into the firebox. He turned to Chippy, who had removed his cap to scratch his head in bewilderment. There was an eternity of silence as the two men faced each other, then the foreman spoke: 'You want to take more water with it, Chippy, or pack it in altogether.'

Of course, we couldn't use the 'Big Yank' again, but a heaven-sent chance of revenge came Chippy's way the following morning. At ten past six, he issued all the gangs of cleaners with their first engine numbers and stores requisitions. There was a distinct gleam in his eyes as he gave us 6647, a 5600 class 0–6–2 tanky. There was nothing unusual about her appearance. No dirtier than the average loco after a day's work, so what was the wily old boy up to? As usual, we took an hour over a two-hour job, then spent the bonus hour on some form of mischief before going to Chippy for our next engine. His expression of joyous serenity remained unchanged as he handed us the stores slip.

Our senior hand read the number aloud: '6647? We've just done her, Chippy!'

Chippy's smile broadened. 'I know that, son, do her again.'

We cleaned her four times during the shift, and the following morning the ritual started all over again.

By midday the shed was rife with rumour, as fitters arrived to give her a more than average check-over. That Churchill had commandeered her as his private loco was a hot favourite.

After three days she would have outshone any inmate of a railway museum. Veteran Barry Railway drivers paused to admire her gleaming beauty with such remarks as: 'Well, I've never seen a Western engine turned out that smart before.' 'Tallow wave the tanks,' said another. 'That's what they did on the Old Barry.'

Chippy revelled in the situation, for the engine was so much in the limelight that we dare not leave her. He constantly appeared to check our work; even venturing down to the drop pit to run his hand along the mirror-clean motions, while the wag of our gang queried if Churchill would do likewise.

The fire-lighter arrived on the fourth day, his usual spartan appearance enhanced with a brand-new suit of overalls and I swear he dusted each lump of coal before placing it in the firebox, but the climax of the mystery was the arrival of the fireman and driver. They walked towards us in the sombre splendour of knife-edge pressed overalls, with polished boots and collars and ties. The two senior footplatemen of the shed, with holier-than-thou expressions, boarded the humble tanky with the dignity reserved for a 'King' class.

Two hours later, as dusk closed in on our afternoon shift, she quietly left the shed to a bewildered audience of workers, who doggedly hung on to their individual versions of the nature of her mission.

With the current spate of wartime films embossed on our minds, we cleaners were most romantic in our judgement. We were positive that we saw the driver carrying sealed orders!

The following day all was revealed, as 6647 returned to the shed, only to be dispatched within the hour on a valley coal train. A dramatic reversal of her regal mission of the previous night, for no one had connected her meticulous preparation with a Royal visit to South Wales, however menial her task.

It was many a long day before she eventually lost her nickname of 'Royal Radiator' – for her duty was to heat the Royal train the night it was shunted into the mile-long Wenvoe Tunnel as a precaution against air-raids.

There is little doubt that she kept the King and Queen warm in their beds, but there is doubt that those two good people were in the least concerned about her appearance. Chippy, however, was adamant whenever we broached the subject. 'You never know,' he said with a wry smile. 'He might 'ave said to 'er "Come on, missus, let's have a stroll along the four foot an' see what's up front." '

2 · Nightriders

The call-boy – that was the engine cleaner's role during his night shift. A pathetic little character who would enhance any Dickens novel. A frightened, grimy-faced youth, who trudged the blacked-out streets from midnight to dawn calling out the footplatemen of his shed. His only companion was the hand oil-lamp to light the way and, for the more elite, an alarm clock to rouse him from his slumbers between calls in shop doorways or park benches.

One such person was Len Tobin. Poor Len. Nothing ever went quite right for him.

I recall the night we both cycled toward the shed after seeing the horror film *The Picture of Dorian Gray*. 'Creepy, wasn't it?' said Len. I tried to erase the most frightening scenes from my mind in anticipation of the dark streets beyond, but Len was persistent in his unintentional scare campaign. 'You couldn't have a worse call than yours after seeing that lot, could you?'

Mine was known as the 'boneyard' call, because it was centred on the local cemetery – a perfect setting for my imaginative mind to recreate the film.

The horrors of the outside world, however, soon vanished in the warm, cosy atmosphere of our messroom, where we gossiped and drank steaming hot cocoa from our enamel tea-cans. I had found a set of oil cycle-lamps in my grandfather's rummage chest during the day – a welcome find in a torch battery-starved Britain during wartime. The rear lamp, we discovered, leaked a little but should last the night on one filling. Len had also been successful in purchasing a set of substitute rubber brake-blocks for his cycle. 'Bloke in the shop reckons these blocks will pull me up just as good as the rubber ones,' he said. Suddenly an explosion as loud as a shotgun filled the air, followed by a cloudburst of baked beans in tomato sauce. Len had forgotten to pierce the can as it cooked on the red-hot messroom stove!

At eleven-thirty, Mr Chidgy, the shift foreman, appeared in his grey dustcoat and black bowler. His target was Tobin. There was a deathly hush, as he made his habitual approach to the rear of his victim to place his hands on his shoulder.

'Jack Daniels, mister.' He addressed everyone as mister. 'You haven't called him once this week. Been late every morning he 'ave.'

Tobin knew better than to argue with a Great Western shed foreman. They were invincible, a law unto themselves – Godhead – and for a mere cleaner to dispute the word of a driver was blasphemy. Mr Chidgy passed sentence with the slight trace of a suppressed smile. 'If he don't turn up tomorrow, me lad, it's up the path for you and no mistake.'

It was an hour later that we six call-boys cycled along the infamous path to begin our duties.

Suddenly, as we went our different ways on a coal-black moonless night, I became very lonely and just slightly frightened. I started a nervous whistle, as I cycled around the town cemetery to make my first call. I welcomed the intrusion of a single German bomber to my ghostly world, as the air-raid siren began its nightly mournful wail. A few searchlights probed the sky, as the distant throb of the German engine crossed the Bristol Channel to lay mines in that busy sea lane – then all went quiet, and I was left with the cemetery and my imagination. I was unaware that my ageing rear tyre had decided to split and allow a bubblegum balloon of inner tube through its smiling crack.

As I cycled on past the vague outlines of grotesque tombstones, Dorian Gray appeared to dart between them, more gruesome than ever Hollywood depicted – and then the inner tube bubble burst! My ailing courage vanished, as I cycled furiously, and with complete disregard for my tyre, towards the town bakehouse; not for the usual welcome warmth of the ovens and the taste of a hot rock cake, but the company of a fellow human being in the shape of Tobin, who always met me there.

Tobin's face beamed, as we met beside the ovens. 'Look at this,' he said. From his overcoat pocket he produced a brass door knocker. His normally passive eyes sparkled. 'From Jack Daniels's front door. I borrowed a screwdriver from the stores after Chidgy told me off. Daniels answered my knock, as usual, then I nicked his knocker to prove I'd been there.'

After a few rock cakes and a cup of tea from the friendly baker we were ready for work again, and I was in trouble for a missed call if the driver near the cemetery complained.

The cunning genius of Tobin prevailed, however. 'Well, you couldn't carry on with a tyre like that,' he said, 'so you met me down here in the town and I fixed you up with a spare wheel I had at home.'

Later, as we changed wheels in the cosy comfort of his neat little terraced house, we discussed Dorian Gray once more. 'I saw the blighter everywhere tonight, I did,' said Tobin. 'Wish I hadn't seen that blinkin' film. Makes you feel creepy when you're out there in the pitch dark, don't it?'

I agreed eagerly with his sentiments.

The call-boy's lot was quite lonely and rather fearful, even on an average night, but this night was blackout-black and Dorian Gray reigned supreme in our young, imaginative minds. We were spontaneous in our decision. 'Let's work it together,' we said as one.

Later, as a single searchlight tried to pick out another lone

German bomber, I returned to my 'boneyard' call with the courage of ten men, having Tobin for company, of course, but the idyllic situation for us was not practical for his slumbering clients. 'I'll have to get back down the town,' he said, and once more I was left to my own little world of self-inflicted misery, as he sped off down the steep College Hill, shouting 'I'll see you at the bakehouse at four o'clock.'

The air-raid intensified during my hour of solitude, and, by the time of my last call, I had the company of half a dozen searchlights, together with the spasmodic gunfire of the local anti-aircraft regiment and, of course, the solitary German raider.

As I neared College Hill, I was unaware that the tail of my railway overcoat had been absorbing the paraffin from my leaking rear lamp, and as I hurtled down that gradient with a zest for more hot rock cakes, tea and company, the draught caused my ancient lamp to burst into flames and ignite me. I was half-way down the hill when a noise, similar to a large blow-lamp, caused me to look behind where my burning overcoat was fanned by my speed. In such a black-out-conscious society, I was more concerned about the tell-tale light than the damage it might have caused to me. However, I dismounted quickly, disrobed and extinguished the offending flames to continue my journey minus a third of my Great Western overcoat.

At the bottom of the hill, however, near an unlit cast-iron Victorian lamp, another light – a torch-lamp – pierced the jet-black sky. As I neared the spot, I could hear groans. It was Tobin. He had collided with the lamp-post at the road junction which it lit in more peaceful times. Poor Tobin. He was semi-conscious, with a nasty gash on his cheekbone. I managed to straddle him across my shoulder and headed for the nearest house. The young woman who answered gave me a terrified look, then broke into a scream, as she slammed the door in our face.

It was only about three weeks later that she redeemed her unchristian attitude when her neighbour, who was a railwayman, informed her of me. Like countless thousands of women at that time with husbands in the forces, she was alone in the house. She was also of a nervous disposition, and that night she, too, had gone to the local cinema. 'When I opened the door', she said, 'and saw this scruffy-looking character smelling of burning rags and carrying a blood-dripping body on his shoulders, I nearly fainted with fear, after seeing that Dorian Gray film as well.'

I took Tobin to the nearest Home Guard post, where they arranged to take him to hospital and where he soon recovered.

The drama continued back at the shed, where I felt honour-bound to present Jack Daniels with his door-knocker when he arrived late complaining of no call.

But there was a humorous ending to a rather gory night, for I discovered that I was not alone on College Hill. Engine driver Ruben Jones, a strong Welsh chapel-goer, arrived breathless at the shed. When his fireman asked what was wrong, as they prepared their engine, he replied, 'I know you won't believe me, mate, but a great ball of fire passed me on College Hill as I was coming to work.' The fireman grinned as the driver summed it up: 'I tell you, mate, it was either a new Jerry invention, or the devil himself!'

Call-boy stories could fill a book, and many were of a frightening nature. I recall knocking an engine driver's door with no response; he was obviously sleeping like a log. Suddenly the front door flew open with an ear-piercing scream to reveal a hideous character. My legs wouldn't move fast enough as I headed for the shed. I learned later that the apparition was the driver's high-spirited twelve-year-old daughter, whose favourite pastime was scaring the pants off call-boys.

On other occasions we would conspire in a group to

frighten a single individual, such as the incident of 'Keris daughter'.

A passenger fireman lived in a railway cottage in the middle of Porthkerry Woods, a delightful place during daylight, but extremely sinister at night. There is a legend that one of the Norman knights of the submerged Porthkerry Castle had a daughter named Keris, who was a witch. Led by her white hound, she would set out each night on her mysterious journeys, to return at dawn.

Generations of engine cleaners at Barry Shed had held on firmly to the story that the ghost of Keris still walked the wood at night, faithfully preceded by her dog. Many a call-boy was scared to death by his colleagues before he even set out from the shed, but a more enterprising band of spine-chillers emerged in my time. Their victim was chosen because of his rather boisterous nature.

The problem was a white hound, or the lack of one. One wag offered the services of his white Jack Russell, which would have looked quite ridiculous preceding the ghost of a Norman princess. Eventually a large shaggy brown creature of every breed imaginable was procured and given a liberal coat of ceiling white powder, to which he offered no objection.

A lot of enginemen missed their calls that night as we all headed for the scene of our crime in the heart of Porthkerry Woods. The night was quite clear with moonlight as the poor lad whistled along his way nervously whilst we sat concealed in bushes and trees to watch the act. Suddenly the ghost of Keris appeared before him in dubious Norman costume. Her white docile hound walked timidly at her side emitting silly little clouds of white dust with each step, but it was all too realistic for the victim. He stopped dead in his track before screaming out 'Christ, it's Keris and her ruddy dog.' Then he threw his oil-lamp at the disguised cleaner before running off into the wood. The lamp exploded in flames to ignite the white robe of the 'ghost' causing her rapidly colour-changing hound to follow the assailant in nervous pursuit more than in anger. That night Porthkerry Woods was a scene of bedlam, as a terrified call-boy, followed by a dusty mongrel, and a smouldering 'ghost' of Keris all made their way back to the shed, whilst five cleaners, hysterical with laughter, completed the procession.

Looking back, however, it was a cruel job to set a young lad to do, and I have yet to meet an engineman who didn't absolutely abhor the call-boy's job.

3 · Italians in Westbury

The greatest moment in the life of an engineman was the day he was 'made'. Strictly on a seniority basis, he was promoted from cleaner to fireman, and fireman to driver. But rarely did he remain at his home depot for his status advance, and a transfer to any of the Great Western sheds usually went in hand with promotion.

Many chose to settle down in the town of their transfer, bringing about a multi-accent atmosphere within their sheds. At Barry, for instance, there worked such men as 'Banbury' Bill, 'Didcot' Davies, and 'Whitland' Williams. Others, of course, applied for a return to their home depot as soon as possible for various reasons – such as severance from their way of life, which also included their work-style – for Great Western sheds varied immensely.

There were some very small rural areas where the shed housed a single 48xx class tanky for push-and-pull passenger work. Here the enginemen carried out most of the locomotive husbandry, and their way of life was poles apart from the busy mixed sheds, such as Barry, or the exclusive main-line depots, such as Westbury.

It was to Westbury's fine shed that I was 'made' fireman in March 1944, where I stayed for some eight months before returning to Barry. Here I experienced my first taste of a spare link, spending hours of boredom in the messroom, whilst my driver, George Blatcher played dominoes. Unfortunately we were not allowed to leave the cabin in case the foreman required us for an urgent job. We had done precisely nothing for about half a shift one night when he approached my mate with orders to turn a loco.

Engine numbers meant little to me in those early days, and when George, too engrossed in his dominoes, delegated me to do the job, I rushed off, with pictures of Barry Shed in my mind, to move a little tanky. Little did I expect that the engine would be a 'King'. I approached the loco from the tender and she seemed to go on for ever as I looked along her boiler, while climbing the footplate steps. With little thought of the Rule Book – and a two-man crew – I blew off the vacuum brake and unscrewed the hand brake.

With only 100 lb of pressure on the clock she was slow to respond to the regulator, so I gave her more. Suddenly, she leaped forward towards the turntable and I braked her in fright. I eventually did complete the 30-yard trip with several stops, only to realise that I had never used a turntable before. At Barry it involved a three-mile journey around the shed with several point changes, ironically through Low Level, where Woodham's Yard is now situated. I didn't realise that the loco had to be balanced on a turntable. I began to wish for Barry and Low Level after numerous attempts at trying to push an immovable loco, and 'King' was the last name for her in my vocabulary as I struggled in the murky bowels of Westbury Shed. It was a pathetic and embarrassed young lad who interrupted the dominoes game much later to admit that he couldn't balance the reluctant 'King'. George Blatcher, a heavily built Wiltshire man, threw down his dominoes in disgust, shouting, as he stormed from the room, 'Bloody kids from tanky depots. They're bloody lost in a proper engine shed.' He was by far the most difficult driver I ever worked with, taking a delight in my natural incompetence.

My first footplate encounter with George took place on an old 'Aberdare' 2–6–0 freight engine which we took over at Westbury heading a parcels train for Bristol. The fire wasn't too clean, it was my first tender engine working, and it was precisely the tenth firing turn of my career, not to mention my first mainline journey. As we pulled away, George began his ridicule, by drawing a chalk line across the centre of the footplate. 'Now that's your side, an' this is mine, and I don't want you or any coal over here.' The fire just wouldn't liven up. Both dampers had been closed by the previous firemen during the waiting period at Westbury sidings, and the clinker had hardened up.

It wasn't a heavy train, and once the engine had it underway we seemed to cruise along on the small valve without effort. But, nevertheless, the blower was hard on, and both dampers were wide open to get a draught under that solid block of unbreakable clinker. More by luck than judgement I managed to keep the needle on the mark and the water above the top nut, then as we hit a sharp bend I nearly departed from the footplate, but not before my shovel of coal landed at George's feet. He said nothing, but his cold silent sarcasm cut me like a whiplash. Going to his short jacket he produced a length of strong cord from its pocket. 'Here, mate, better tie that across the gangway, you might not be so lucky next time.' It was a tired and depressed young fireman who made his way home that night to his most unwelcome digs.

Because of the massive amount of rail traffic building up to the D-Day invasion, and the strategic position of Westbury, the capacity of the shed had been doubled in engines and workforce. Consequently, all the usual homely lodgings of that town had been well and truly saturated, and my billet for six months was in the infamous 'Coaches'. Many enginemen of that period will recall the train of seven sleepers and two restaurant cars that was our home.

Not many will look back with nostalgia upon the Italian prisoners-of-war, who sang beautiful arias as they cleaned the coaches whilst we tried to sleep after night duty; or the constant type of sideways walk that was peculiar to those narrow corridors. The whole life-pattern became slightly claustrophobic on a stationary, yet never ending, train journey.

At the head of this depressing hotel was an old double-framed 'Bulldog' class 4–4–0 tender engine – the type with side-plates covering most of her wheels. Her job was to supply hot water and steam for the miserable inmates behind her, and her caretaking was the responsibility of newly 'made' firemen.

All my dreams of working a 'King' class out of Westbury Shed were dashed to pieces as I watched the Cornish Riviera hurtle past from the footplate of the old 'Bulldog', where I did my stint as coach warmer.

Memories of twelve-hour shifts spent on that motionless old loco still make me wince, as I recall the few shovels of coal each hour, and the pleasing song of the water injector to break the monotony – but there were moments of interest.

One of the Italian prisoners befriended me. A fine chap, with a deep interest in my old 'Bulldog'. We spent many hours of pleasant company with each other comparing our countries and our customs. He spoke with a genuine concern of the futility of war, as the countless American troop trains changed crews at Westbury, while my youthful mind dreamed of the day that I would wear the uniform of my country but, of course, the opportunity never arose, for we were employed on a high-rated reserved occupation.

I was fortunate to miss my caretaker job on the old 'Bulldog' for two weeks and enjoy the thrill of working on the moving engines of various goods trains to the South Coast and West Country, but, when I rejoined the old loco, the drama was as intense as any mainline episode.

I relieved a worried and tired fireman, as he hurriedly packed his bag to leave the footplate. 'She haven't been on the shed for a week, mate. Look at her!' With the blower full on, he opened the firebox doors to a lifeless fire. 'Clinker right up to the ring, mate, and look at the clock!'

The steam-pressure gauge needle showed just on 100 lb, whilst the boiler-water gauge glass was barely a third full. Before I had time to comment, he had gone, leaving me in the most ludicrous situation imaginable.

Even in my early career, I had already suffered the shameful, ultimate defeat of 'stuck for steam'. At least I looked the part, as my grinding loco came to a wheezing halt, while struggling to haul sixty coal empties up a steep Rhondda Valley gradient – but to suffer the same ingredients of those nail-biting moments on a locomotive that never moved was beyond belief!

As the steam gauge continued to drop, and the water level headed for the bottom nut of the glass, I dropped everything and ran for the nearest line-side phone box.

The lordly voice of the foreman answered my plea for a driver to take me on the shed for a fire clean: 'Have you got enough coal, lad?'

'Yes, foreman.'

'And you have a water-hydrant leading into your tank?'

'Yes, foreman.'

'And you have fire-cleaning tools on board?'

'Yes, foreman.'

'Well, why the hell are you bothering me? Clean the fire and don't forget to spread the clinker along the four-foot when it cools.'

I had often shunned the idea of ever tackling a normal fire-cleaning operation, but the 'Bulldog' must have contained the dirtiest fire in the history of railways – a sea of clinker, at least a foot deep, covered the whole of her box. I made a dive at it with the long chisel bar, only to vibrate throughout my body as if I had done the same to a granite rock face. At that moment I decided that I wanted to terminate my railway career forthwith, and, as I pondered on the subject that fire-droppers must be Herculean, sub-human creatures of another species, a jug of tea appeared on the footplate.

'Ah, my good Taffy friend! I hears you are back and I makes you a jug of tea. Good! Yes?'

It was my Italian friend – Carlos. He quickly assessed the situation, as his eyes travelled from the steam-pressure gauge to the water glass and the lifeless fire.

'Ah, you got troubles! Yes?'

'I can't make an impression on it,' I said.

He took off his brown battle-dress blouse, and shirt beneath, to reveal a superb torso, and to intensify my feelings of inferiority he sang as he began his attack on my monument of clinker. Carlos certainly knew how to handle a fire, as he hacked, chiselled and shovelled one half of the box clean to start a fresh fire on the clean bars – then start on the other side.

It was well over an hour, and countless tenor arias, before the 'Bulldog' could boast a set of clean bars, and, as the steam gauge slowly crept up to a respectable position whilst the injector just managed to keep the water in sight, my Italian hero, dripping in sweat, gave me a broad smile as he sank back onto the driver's seat.

'Some hell of a dirty fire you had there, Taffy!' he said, as he donned his prisoner garb.

I felt forever indebted to Carlos, and our friendship grew stronger as I managed to provide him with some of the luxuries that were in short supply to him, such as cigarettes and chocolate.

During a weekend leave to my home, I attended the local church bazaar and, of all prizes, I won a bottle of Chianti on the bottle stall!

Carlos appeared depressed when I gave it to him two weeks later on the footplate of the 'Bulldog'.

'What's got into you?' I said. 'I thought a bottle of that would really have relieved your home-sickness.'

He shrugged his shoulders. 'Then you haven't heard of my trouble with Bully Harris?'

The man he mentioned was a mainline goods fireman. A loud-mouthed character, who had earned his name well by his aggressive attitude. He also imbibed rather heavily in the local Scrumpy cider, being a thorough nuisance in the coaches after a 'session'.

'How did you cross him?' I said.

Carlos half-smiled. 'You know how I like to sing – well, I was cleaning Bully's window. I thought he was on days, but he had changed shifts – he was on nights. It was a lovely day and the birds, they were singing. I know I been warned, Taffy, but I couldn't help it, Taffy. I joined the birds and I sang. And Bully, he wake up mad. He come round and pull me off the ladder an' we fight like crazy.'

I could well imagine that Harris had met his match that day, and I wasn't surprised when Carlos told me how Bully had reported him to a sergeant for assault.

'I just been let out,' said Carlos. 'Two weeks confined to camp and no privileges. I tell you, Taffy, that Bully Harris, him a bad man!'

But the trials of Carlos were not to end there. Bully, taking advantage of his superior position over a prisoner-of-war, persecuted the poor Italian to the limit. Two weeks later on a Saturday evening when most of the coach dwellers had gone home on week-end leave, I sat in the restaurant-car on my own. Carlos, who was on duty, approached me.

'Taffy, there's only you and Bully sleeping in the first coach tonight. You do me a favour, please. You take a bed in the second coach. There's plenty to spare.'

I guessed he was after some form of revenge, and I nodded agreement.

Later that evening, Bully staggered into the dining-car

very much under the influence of Scrumpy. Fortunately he could barely stand, and soon he headed along the corridor to his bunk. As I lay on my pillow in the second coach, I tried to solve what fate might lie in store for him, but sleep soon took over until dawn, when I was awakened by the excited chatter of a number of Italian prisoners. It appeared that there was no steam or hot water available. In fact, there was no locomotive! She had vanished with the first coach. They were discovered eventually about mid-morning in a siding four miles away, with only Bully aboard, sleeping as sound as a log.

Try as he might, his pleas of innocence fell on the deaf ears of the head foreman, a staunch chapel member, who insisted that the 'demon drink' had caused Bully to steal an engine and coach for joy riding. He could have been dismissed from the Company in peacetime, but a registered caution was his sentence, and a few weeks later he was thankfully posted to another depot.

Back on the old 'Bulldog' some weeks later, I met Carlos again. Recalling the expert way that he handled my dirty fire, and bursting with a desire to know how Bully, the 'Bulldog' and the first coach landed up in a siding four miles away, I said, 'You seem to know quite a bit about railways, Carlos?'

His perfect white teeth gleamed, as he broke into a broad smile. 'I was an express fireman, based at Turin, Taffy, but only you and I know that.'

4 · The Shedmen

Barry running shed, although built by the Barry Railway Company before the turn of the century, was quite similar to the design of many GWR sheds. It had no such luxury as a turntable, and any visiting tender loco had to make the three-mile trip around Low Level to turn.

There were five through roads, and one dead end, where the fitters' quarters were situated. These must have been built as an afterthought, and whilst boards covered the pit, no stop-blocks protected the building at each end. The situation was the subject of a long-standing joke based on the American song 'The Railroad Runs Through the Middle of the House'.

Over the years there had been some near misses and some not so near, when the shed berthers, constantly moving engines in low steam, failed to stop outside the building. The accidents were always slight, but nevertheless the experience could be quite hair-raising when the buffers of a loco came through one of the walls.

The 'spare' engine, always a 5600 class 0–6–2 tanky was berthed 'alongside the shed'. Invariably she would work, or not work, about eight hours of spare duty before being booked on a job. On rare occasions some shed foreman forgot to change her for a day or so, and consequently she would go out on a job with the start of a dirty fire, which could mean a hard day's work on the fireman's part even for a '56'.

The coal stage was another area to suffer the ravages of marauding locos. The man-made incline up to its first floor was extremely severe, and a '56' certainly barked a bit to hump eight 20-tonners up its bank. The trouble was the lack of room. The actual stage housed four or five wagons, leaving room the other side for the remaining three or four against a stop-block. No loco was allowed onto the stage so we had to stock up with eight wagons. When the coalers had emptied their first four wagons they would release the brakes of the little train to allow the next four wagons in. This exercise should have been carried out, of course, by a loco, but on most occasions the coalers didn't bother to ask for one. There were times when the brakes were not too good, and eight wagons ended up in Low Level sidings, if there was nothing in the way beforehand. Conversely, with so little margin for error at the top end of the stage, combined with the efforts of the hard-hitting loco, the stop-block was often stretched a little.

Number 1 road was the boiler wash and tube cleaning road, where an overhead steam pipe carried adaptors for the tube cleaning lances. Another excellent purpose for the lances was to free frozen points during the winter. Normally a linesman would clear the snow or ice from the points with his spatula and spread some salt over the runners, but on extremely severe occasions, when he was unable to cope, a loco and a steam lance – strictly against the rules – would carry out a perfect job, but without the salt.

There was said to be the ghost of an engine cleaner haunting No. 2 road. Apparently the poor lad hadn't changed his overalls in weeks and they were saturated in oil. It was just before the First World War in the days of Barry Railway, when one boy cleaned one engine throughout a shift. And clean it he did, with tallow, Brasso, and best cleaning oil. The regular driver wouldn't hesitate to report the matter if he so much as blemished his overalls. The young cleaner was working a night shift with his ducklamp for illumination, but unfortunately all the shed doors were open at both ends allowing a keen wind to blow through the building. The theory was that the ducklamp ignited his oil-sodden overalls, for his colleagues saw him, completely enveloped in flames, running toward the top of the shed and possibly the water column. He collapsed half-way along the shed, and died a few days later. Ironically he could have jumped into No. 1 road pit, which always contained a few inches of water from the boiler-washing activities.

Steam raising had many facets. The usual manner for a cold engine was the use of 'sticks' and oily waste. A 'stick' comprised twelve ordinary household sticks nailed together criss-cross to allow the oiled waste to be inserted between. About four of these were placed just under the firebox door with a few lumps of large coal.

The steam raiser would have about twenty locos to light and attend during his shift, and six hours was the desired time for a cold loco to reach a full head of steam when the fireman had prepared her.

If the loco was required very quickly, the firedroppers would be instructed to 'save' a fire. To do that they would clean one half of the box, then push or shovel as much clean fire as possible over to the clean bars, and finally clean the other side. The fireman would always recognise this type of fire when he signed on, as it stretched the whole length of the box on one side.

The quickest way of raising steam on a cold engine entailed the use of the scoop, but many enginemen stressed that it was harmful to the firebox and tubes. The scoop was a huge rounded shovel with a long steel handle. It would hold over a half-hundredweight of fire taken from the drop pits or the sand drying fire. It was a terribly awkward tool to use and uncomfortably hot for the person who carried it. It always caught the wooden footplate alight (at least whenever I used it), but it did move the steam-pressure gauge round quite quickly.

On reflection, I think my fondest memories come from the shed. A dirty black, draughty, smoky building, but bursting with character – an old broody hen to her busy little chicks, the locos. A haven to return to after a bad day.

When one considers the small army of staff who carried out the valet service on a steam locomotive during her period on the shed, it is quite understandable that her much-maligned successor, the diesel, was an obvious choice of economic necessity.

Let us pass through the steam into the past and take a look at a loco as she arrived on the shed after completing her turn of duty whether a shunter which had been out for twenty-four hours, or a passenger engine which had emptied her bunker in less than eight hours.

As soon as the enginemen left their charge, 'Smokebox Charlie' would emerge from his wheel-less guard's van dugout. His sole duty was to shovel the fine black ash out of the well of the smokebox. One of these chaps, Jack Connors, habitually carried a small bag of the ash home each day. 'Slugs,' he would say when you passed him, 'there's not a slug that can live with it on the garden. It's so sharp, see! Cuts their bellies to pieces.'

Next came the Top End shunters – a fireman and driver, usually 'shedded' on medical grounds, who moved the loco on to the drop pit for firebox and ashpan cleaning. When that was completed, two more enginemen, the 'berthers', took her on to the coal stage for fuelling.

Then it was the turn of the Bottom End shunters – three more enginemen who moved her into the shed. But that attention was only the start. In their turn, with meticulous timing, came the boiler wash crew, the tube cleaners, the engine cleaners, the fitters, the steam raiser, and, finally, her fresh crew, who would start the work cycle all over again.

Ben Turk, a tube cleaner, was a strange old man. At least he appeared old when covered from head to foot in the light grey dust of the firebox, where he spent most of his time pushing his long steel rods through the boiler tubes. In the subdued light of the shed his appearance was ghostly and could be quite frightening when one was unexpectedly confronted by him. He was also deaf, which was a serious disability for any shedman, for he never heard the warning shout of engine crews who were about to move the loco. But, as his place of work was the firebox, he was in no real danger in comparison to fitters or engine cleaners, who could be working on the motions, and there personnel usually placed a 'Not to be moved' sign at each end of the loco.

The story goes that a 5600 class tanky had to be pulled to the engine fitting shop, about a mile away, for her general overhaul. A driver and two firemen, with the aid of a 5700 class pannier tanky, shouted warnings all around the '56', then coupled up, while one of the firemen took her handbrake off, and for no particular reason, closed the firebox doors.

An hour or so later, two mischievous fitting shop apprentices, who were evading the intentions of their tutors, decided to hide on the footplate of the '56'. As they puffed away on their cigarettes, a tapping sound came from within the firebox. The boys had been told of boiler ghosts during their initiations – the greyish-white apparitions of old boiler-makers, who had been accidentally riveted in their tombs when the loco was built. Gingerly, the bravest one slowly pulled the door lever to reveal the ghostly face of Ben Turk. The youths didn't wait to hear Ben's story, and, to this day, they may still talk of the boiler ghost they saw when they were apprentices.

I remember how I scared the life out of a shed foreman in complete innocence. I was one of the Bottom End shunter team on a rather dirty, drizzly night shift. I was elected to be the 'bookman', my duty being to walk from the coal-stage to the drop pit, taking the engine numbers in rotation. Then I would go into the foreman's office for their movement instructions. Mr Chidgy was the foreman, and I noticed that not once did he look up at me as I called out the engine numbers that he entered in the rota sheet.

The miserable night wore on, and, during the half-way mark, I very lightly knocked my head, just above the eye, on the steam-heating pipe of the loco that I was uncoupling. She was the last engine of that particular batch, and I began my walk to the drop pit once more. The misty rain seemed to saturate my face, and I constantly wiped it with a piece of cotton waste. Finally I reached the foreman's office. He acknowledged my entry with a grunt, not taking his eyes from the rota, which he always studied intensely. The last engine on my list was a 'foreigner' – one from another shed.

As soon as I said the number, he looked up at me, his face taking on a feature of terror. 'Oh, my gawd! Oh, my gawd!' he repeated, before slumping unconscious in his chair.

At that moment, Nick Williams, the time-keeper, burst into the office from his adjoining room; his eyes moving swiftly between me and the foreman. 'What you been up to, lad?' he said in a shaky voice. 'What you done to him? Must have put up a hell of a fight, he must, to mark you like that!'

I started to walk towards him. 'What do you mean, Nick?' I said. 'I just gave him an engine number and he looked up, then he fainted.'

In complete bewilderment, I walked into the foreman's toilet room. My body stiffened at the sight of my blood-stained face. I filled the wash basin and washed the gruesome evidence away, leaving a small cut above my left eye. It was the injury caused by the heater pipe earlier on, and I had assumed the dampness on my face to be drizzly rain. Of course, I had worsened my appearance by constantly wiping it with the cotton waste.

I was known thereafter as the 'foreman killer', along with my other title of 'Admiral', due to my ownership of a small rowing boat.

The 'sand dryer' was a member of the shed rarely mentioned. His job was to stoke the fire that baked the coarse sea sand until it ran like liquid through the sand pipes of the locomotives.

Bill Pearce was the sand dryer at Barry Shed. He was also the unofficial first-aid man and purveyor of many strange elixirs of good health. Ordinary lamp paraffin was his favourite cure-all. His breath stank of it, and we were constantly amazed that he could light his pipe without blowing his head off. He was also extremely good at removing the dreaded sharp ash from our eyes, but he loved to dramatise the whole affair into major surgery, by donning a dirty white coat and a specially adapted cloth cap with a magnifying glass screwed to the peak. His operations would usually be accompanied by some onlookers pretending to wipe his brow and offer him the paraffin bottle. It was a strange coincidence, however, that the magic paraffin could never keep Bill in good health whenever a wagon of sand arrived on the scene. He had a constant inexhaustible supply of ailments – ranging from dysentery, caught in the Dardanelles during the First World War, to some rare oddity, such as Blackwater Fever. But a wagon of sand to empty, apart from being sheer hard labour, could be a blessing to any young fireman who desired an early finish during an afternoon turn.

The cunning shed foremen always used the same tactic when Bill went conveniently sick. 'Wagon of sand, lad,' they'd say, picking their victim carefully. 'Empty it and finish.' They were usually 12-ton wagons of wet sea sand to be shovelled 6 feet up into the dryer.

How on earth we had any strength left to socialise, after

displacing that much sand in a matter of hours, is now an eternal mystery to me in my present middle-age.

Perhaps the best story, embracing a complete shift of shedmen, is the saga of *Band of Hope*, an ageing 30-foot ex-ship's lifeboat, owned jointly by twenty one-pound shareholders. Her weakness was her engine – a 4 horse-power generator motor from a Cardiff hotel, hardly super-powered to combat the notorious 7-knot currents of the Bristol Channel. Our merry band set sail one fine June morning for a few hours fishing, prior to their afternoon shift. The little engine, untried beyond the harbour, pushed *Band of Hope* down channel in fine fashion on the slack water of full tide. About midday, when a good bag of fish had been boated, David Baggot, steam raiser and captain, decided to up anchor and prepare to go to work. But the ebb tide was in full flow and the gallant little engine, with full throttle, failed even to slacken the anchor rope.

Back at the shed, an anxious foreman stared at the clock, as it reached two-thirty.

'What the hell do you reckon it is then?' he said to the puzzled time-keeper. 'We're in a right mess. No shed engine-men. No steam-raisers. No fire-droppers. No coal-stage workers. No ruddy nobody for that matter!'

The time-keeper didn't choose his words wisely. 'Don't s'pose it could be a flu epidemic?'

Mr Edwards stormed out of the room, too furious to reply. He had enough spare enginemen to keep the loco-motives on the move, but no men to service them. The awful truth dawned on him when driver Bill Hobbs signed on at four o'clock. Bill was a keen shore fisherman, and he had recognised *Band of Hope* about a half-mile off Porthkerry Bay that morning. 'I reckon they're stuck out there,' he told the foreman. 'That little pop-pop of theirs wouldn't pull the skin off a rice pudding!'

Foreman Edwards began to grin, as he lifted the phone to

speak with the GWR Marine Control.

Back in Porthkerry Bay, the miserable crew of *Band of Hope* sat in sombre silence, interrupted only by spasmodic suggestions of genius from various members. Suddenly the still air was torn by the deep, steamy blast of a ship's siren.

'We're going to be run down. He haven't seen us,' said Dai Baggot, as the ship steamed towards them with a bow-wave indicating full speed. When all seemed lost, the vessel veered off, causing *Band of Hope* to rock wildly in her wake. From the wheel-house of the tug, her skipper shouted sarcastically, 'You want to stick to the railways and leave the Channel to seamen. Take a line and we'll tow you home to your foreman. He's worried to death over you!'

About seven o'clock, the crew of *Band of Hope* made a belated attempt to sign on, but Foreman Edwards was unforgiving. 'Sorry, lads, I can manage without you for three hours. You'll have to lose a day. Go and sell your fish to make it up.'

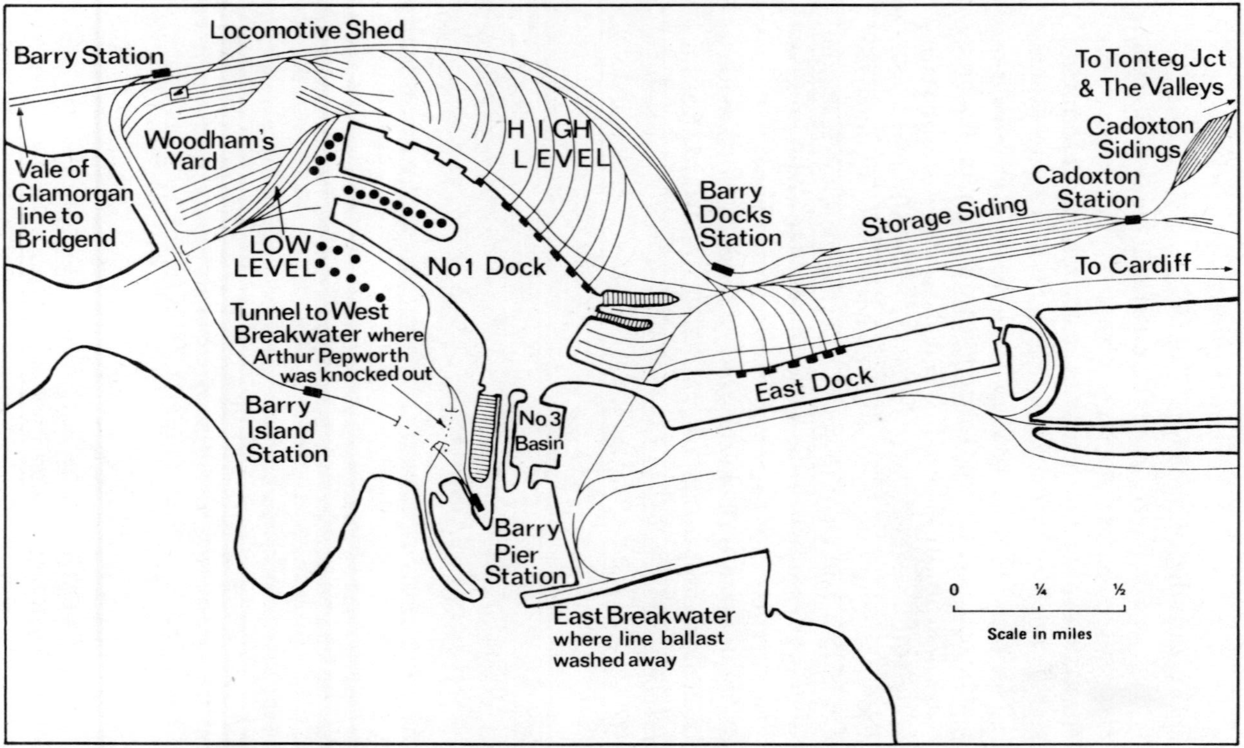

Detail of Barry Docks

5 · Barry Coal Pilots

During its heyday of 1913 Barry Docks held the world record for its coal exports. Originally completed about 1890, the thirty-one coal hoists around the perimeter of over 100 acres of dock loaded Welsh coal to every corner of the earth. Eleven million tons was the figure, and when we consider the size of the wagons (10-ton wooden trucks), the annual movement adds up to well over a million truck-loads. In fact the whole Barry Railway system covered only 68 miles of line, whilst Barry Shed housed 138 locos in 1913, nearly two engines to the mile.

'King Coal' ruled at Barry until well after the Second World War, when steamers still waited their turn to berth under the never-ceasing hoists. It was mainly Rhondda and Rhymney valley coal that came to Barry, hauled in later days by the famous 5600 class. Cadoxton Yard received a constant flow into its eighteen down sidings, whilst the same number of up sidings housed the empties.

Now began the legend of the Barry Dock coal pilots, quite unlike any picture that a railway buff may create of a busy little tanky hauling a few mixed goods. These Barry pilots *worked*, and many was the time that a bunker of coal would last the engine only one eight-hour shift. But the legend was more than the locos, it was the men. Not just enginemen. It was teamwork throughout in days when industrial action was unheard of. The hungry holds of the ships demanded coal, and quickly, in order to sail on the tide, and I have never heard of any failure to respond to that demand.

Imagine thirty-one cargo ships with gaping holds taking a wagon of coal every four minutes. The Rule Book is completely ignored, it has to be, or the whole thing will grind to a halt. A '56' steams down the last incline into Cadoxton Yard. The brake blocks of her first dozen wagons are red hot. The brake van which was uncoupled some way back approaches level to the loco along another inclined siding with the guard in control. Nature designed this clever anti-Rule Book incline movement, and in no time the '56' collects her van and proceeds to the shed. But the train of coal barely stops moving before a busy Cadoxton pilot couples on to it. The grade three shunter walks one side of its length while the grade one walks the other, and with a tail lamp fitted they both return to the footplate. These were happy men with their own individual methods of train working, and no two characters alike.

The Mills brothers (no relationship to each other) could keep you in fits of laughter throughout the shift. 'Six twenties of very large lumps for High Level, driver, and turn the gas up a bit, fireman, 'cos they's going to need some humping!' With the steam gauge fast approaching the red mark the driver blows his whistle for High Level: one crow, one long.

The disc comes off and we're away to feed another hungry ship. During such trips I often mused on the point that the coal we hauled was a permanent feature of this country just twelve hours previously, but in another twelve hours it could be on the high seas.

There were, of course, many humorous incidents on the Barry pilot job. For example, as the driver closes the full regulator, the wheezing, groaning ex-Rhymney Railway tanky comes to a welcome halt outside the two-storey traffic inspector's cabin at High Level. The ancient old lady had good cause to display the labours of her task, for behind her is the long black snake of fifty 20-tonners of Welsh large. The journey was short – less than three miles – but it was a full mainline train, and traffic inspectors, renowned for their lack of compassion towards locomotives, would not make it less for the suffering dock pilot. As the fireman replenishes the thirsty boiler with the two slow running Peacock injectors, the head shunter walks to the traffic cabin for orders.

On the top floor of the traffic cabin, Inspector Light runs his index finger along the lines of the ship's cargo requirements, while the grade one shunter awaits his orders.

'Number six tip,' barks the clipped voice of the little inspector. 'The ship needs about thirty to finish, but put 'em all on number six for the time being.'

Shunter Mills rushes back to the pilot with a twinkle in his eyes. It is Saturday afternoon. If all the coal hoists are serviced, Inspector Light might send the engines home to the shed, and he could enjoy a pleasant evening in the Railway Club. The tanky struggles away with the final haul of her stubborn train. Thirty minutes later, the light engine approaches the traffic cabin – her mission completed.

'Nip up to "Lighty" for the orders,' says Bill Mills to the fireman. 'You're younger than me.'

Taking the steps two at a time, the nimble fireman bounces into the cabin. 'Orders for C6, inspector?'

'Pack up,' says the little man. 'Pack up.'

With unexpected visions of a night out at the local dance hall, the young fireman speeds towards the pilot with the good news.

'I knew it,' says Bill Mills. 'I looked over his shoulder. There's nothing more doing after number six is filled. Come on! Let's make for the shed before the little blighter changes his mind.'

Within half an hour the foursome sit around their first pints in the Railway Club. The fireman rises to his feet, as he completes his drink in a thirsty manner. 'Ah well, boys, that's my lot for the moment. I'll pop up the hill an' get changed. Don't look so depressed, Bill. I'll buy a round when I come back.'

'Not tonight you won't, lad.' The High Level traffic foreman stands in the doorway behind them.

'What's the matter?' says Bill Mills. 'Lighty changed his mind, 'ave he?'

'No, Bill,' says the grinning foreman. 'You should have come for the orders yourself. You know how he talks. He told the fireman "Back up". Back up to Cadoxton for another train. Not "pack up".'

He produces a halfcrown from his waistcoat pocket. 'S'pose I'd better have one now that I'm here. Don't be long now, lads. Better go and grab your engine 'fore they drops her fire.'

But the old Taffy has been dropped. Not a glimmer in her box. Forty pounds on the clock and blocked in by three more locos.

'What the hell we going to do now then?' says Driver Francis to his three colleagues in crime, as they hold conference on the footplate of the tepid loco.

'Got it!' says Bill Mills, with a broad smile of triumph. 'Let's wait at the shed top. There's mainliners and passenger locos coming in all the time. We'll tell the crew of our little problem, and take their engine.'

The other three nod their agreement, and within the hour they are back at Cadoxton Yard on the footplate of a 5100 class passenger loco. What the quartet didn't know, as their unusual pilot puffed around the dockside, was the hue and cry on the shed – 5160 was re-booked for the last passenger of the day as soon as her crew signed off. She should have had a quick fire clean, and bunkering, to be placed ready for her next crew. But she could not be found. They searched the shed from end to end.

'What the devil are you on about?' said the foreman. 'She's big enough, ain't she? Haven't slipped between the sleepers, sort of thing!' He was, of course, completely ignorant of the fact that she had actually not arrived on the shed. After twenty minutes of fruitless searching by all the available shed staff, the luckless foreman decided it was all too much for him. There must have been a confusion of engine numbers. 5160 couldn't have worked on afternoon passenger, and she most certainly wasn't available for the late night passenger. He solved the nagging problem with the provision of a fresh engine number, but later, as he poured through the locomotive work sheet, there was no record of 5160 working elsewhere.

Barry may have been unique in its relationship of enginemen and shunters. The four men certainly worked as a closely knit team, and, for reasons that I don't know, shunters' trucks were never used. Some long-serving shunters, with an interest in locomotives, could well pose as experienced enginemen. I was often indebted to such shunters during my early attempts at firing pilots for, totally against the Rule Book, firemen would change roles with the grade three shunter for a spell on the coupling pole.

The dock pilot work at Barry embraced a brotherhood of men that was certainly unique, a kind of engine work that differed totally from any other. At each yard the centre of activity was the two-storey traffic building. The top housed the office and observation department of the traffic inspector and foreman. Below was the messroom where shunters and enginemen were supposed to take their twenty-minute break, but as long as the work was done, it could go on for an hour or so.

I still have snug memories of that messroom about two o'clock on a cold winter morning. The routine was to take water on the loco, black out the fire at the back of the box, fill the boiler, and knock her right back to around 100 lb pressure on the clock, close both dampers and hope she would stay quiet for an hour. Blowing off was fatal because the inspector would order you back to work right away. The 5700 class panniers were the best; you could switch them on and off like a gas cooker, but the ex-Barrys with their badly worn boilers were another story.

The messroom always had a mammoth fire on the go to boil a giant cast iron black kettle of GWR vintage. There were a dozen or so used matchsticks floating on top 'to absorb the smoky taste', so I was told, but smoke or no smoke, I can't recall a better cup of tea than it produced. Bread and cheese sandwiches seemed to be the perpetual menu, aided, no doubt, by the wartime extra ration of twelve ounces per week. Two or three card schools could be on the go along the 12-foot-long white scrubbed table, and we ate as we played.

It was a favourite swan-song of the older drivers to remind us of a type of tachograph fitted beneath the framing of pre-war locos. Apparently it was to record non-working periods, and would ensure that the twenty-minute break was strictly adhered to.

'Bloody nuisance it was,' they'd say. 'But we had a way out: used to take it in turns driving up and down an empty siding for forty minutes or so, and they scrapped the idea in the end.'

A large area such as the New Dock could have as many as twelve engines working on any single shift mainly to handle the imports.

Quite recently the two remaining coal hoists were demolished, and the Geest Line transferred their banana import activities to Avonmouth. Now only a solitary diesel shunter can be spotted as its wheel flanges grind mournfully against little-used lines of a ghost dock. But the ghost of one particular shunter still lives vividly in my memory each time I visit a particular area of the dock. His name was Arthur

Pepworth, a West Countryman. He and the same locomotive were permanent fixtures of the dock mechanical engineers. 1993 was an open-backed cab 0–6–0 pannier tank, boasting a holed canvas canopy as her only protection against the elements. The job was in the Light Duty link, and the loco rarely pulled more than two wagons and a steam crane while attending to the dock engineers.

Arthur Pepworth was a red-faced little man, who could go several shades redder when his temper rose with the slightest provocation. Most enginemen would humour him, purely for a peaceful life, but to Driver Dick Stonehewer, he was a challenge! Arthur was alien to Dick's light-hearted way of life and from the very start of the week, Dick rubbed him up the wrong way, causing the little man to be permanently crimson, as he ranted and raved. Throughout the week, Dick constantly nodded towards Arthur, when he was out of earshot – 'Ask him who followed the band, mate! Go on! Just ask him that.'

Fearing the wrath of Arthur in response to an obvious practical joke, I remained mute on the subject until the Friday. We were hauling two wagons of granite blocks to repair the west breakwater. To get there we had to go through a single-line, rough-hewn tunnel, about 200 yards long. As we entered, with a short respite from the driving rain outside, Dick came across to my side to whisper, 'Go on mate! Ask him! The week's nearly over. Might be months before you work this job again.'

At last the inquisitive side of my nature won, as I blurted out against the noisy rattle of our old loco, 'Who followed the band, Arthur?'

Were it not for his railway uniform, he could have been mistaken for a Redskin, as he danced before the fire. 'He put you up to it, didn't he? Him there. The wicked devil!'

Dick Stonehewer's eyes streamed with tears of mirth, as the infuriated shunter continued to scream abuse at the both of us. Suddenly he slumped forward, then sank to his knees in a silent daze. We rushed to his aid, with instant feelings of guilt.

'Heart attack,' said Dick, going back to close the regulator and stopping the train outside of the tunnel.

'Heart attack, be damned!' said the hazy voice of Arthur, as he opened his eyes. 'Take more than a couple of loony enginemen to give me a heart attack.' He rubbed the top of his head with a pained expression. 'Ruddy piece of rock off the roof of that blasted tunnel. Expect you'll find it other side of the canopy. If it weren't for that an' my cap, I'd have had a cracked skull, you can be sure!'

I felt around the canvas to find a stone as large as a cricket ball, and with strong feelings of remorse, we made Arthur a can of strong tea, not allowing him to do any further work that day. Later, as I walked home with Dick, I asked him why Arthur 'blew up' when I asked the fateful question.

'Well,' said Dick, 'Arthur used to be the big base drum player in the local Silver Prize Band. Now a church parade was organised one Sunday, and the band had been briefed on the route, but, at the last minute, for some reason or other, the directions were changed. Perhaps, because of his bad temper, no one approached Arthur to inform him of the change.' Dick began to laugh. 'When the band came to the junction of Holton Road and Gladstone Road, little Arthur, behind his ruddy great drum, went one way and the band the other!' Once again the tears began to stream down Dick's face as he continued. 'Everyone clapped and roared with laughter, but Arthur marched on regardless for a good fifty yards before he realised that he was on his own. You should have seen his face, mate, when he lowered his drum to see nothing in front of him. Suddenly he stamped his feet, grasped his drum and ran like hell to join the band!'

Arthur was also a showman who loved to play to the general public. His character changed completely before an audience at level crossings or stations. From his guard's van, just two or three wagons away, he would wave his new crisp green flag, accompanied by an endless variety of signals from his rarely used whistle. These egotistic actions, before a crowded platform one day, were more than a mischievous young fireman could tolerate. Slipping down on the blind side of the platform, he uncoupled the small train from the engine while awaiting the signal. As soon as the board came off, Arthur intensified his act, and the loco pulled away, leaving him stationary before a delighted audience.

Speaking of bands, however, there were firemen who could play quite a tune on the pep pipe*. Some would entwine it into a form of a tight circle and blow one end like a bugle, whilst others would make a noise like a huntsman's horn by ferociously swinging the full length of the pipe in a large circle outside the engine cab as they blew it.

But the scene that once brought tears of laughter to my eyes was outside High Level cabin when a party of firemen and grade three shunters formed an orchestra to recreate Glenn Miller, or so they wildly imagined. Three played the pep pipes on various fashions, one played the upturned buckets. Two blew through their hair combs which were covered in the tissue from their cigarette packets, and finally the nearby water injector of an old Barry tanky was made to sing as the fireman kept her running bone dry. But in no way could my imaginative mind remotely conjure up the sound of 'Moonlight Serenade'.

* The pep pipe was a wire-bound rubber hose about 6 ft long and 1 in. in diameter, which was connected to the pep cock. This supplied a powerful jet of water when the injector was in use, as its primary purpose was to wet the coal and sluice the footplate clean.

6 · Light Duties

There were two 0–6–0 tankies based at Barry Shed, known as the 'Coffee Pots'. Pretty little things, in mint condition considering their age, for they were built during 1894 for the Barry Railway Company passenger service. With only 120 lb boiler pressure, and not very much in the way of tractive effort, there was little use for them, other than their one-coach career of the old Barry days. But in those halcyon days of Great Western management, provision was made for light duties for locomotives and crews, and 783 and 784, ancient ladies as they were, served the purpose admirably.

Light Duties could be prescribed only by Doctor Bennett, the Company Medical Officer based at Swindon, who insisted on examining all drivers and firemen if they had spent more than five weeks on sick leave.

One of the long-term recipients of his administration was Garry Owen, driver of 784, on job number P1, the Ship's Stores. Although his true name was Ralph Richardson, he bore absolutely no resemblance to the famous actor in any manner. The nickname, taken from a newspaper racing tipster, had stuck because of Garry's obsession with that sport. So deep rooted was his interest that he constantly used the hand language of race-course tick-tack to his bemused colleagues, who laughed the matter off, saying 'His trouble's in the head – not the body.'

Beneath Garry's clowning was a shrewd character. The real cause of his long convalescence was a severe crush from the big ends, as he oiled the motion of a passenger loco. She was moved by a fireman who failed to check that all was clear before he went for water.

I was sentenced to six weeks of hilarious recuperation with Garry, after suffering a rather nasty attack of pneumonia. My first encounter with the tipster happened when he threw the shovel from the footplate as I prepared 784 on the shed on the first morning of my convalescence. 'You don't need that for a start,' he said, as he endorsed the statement with a display of tick-tack. 'You won't get through more than two or three hundredweight all day. Put lumps on by hand from the bunker doors – it would be a shame to dirty the footplate.'

As the day wore on, I discovered that the one or two occasional lumps, fired by hand, supplied ample steam to haul the single van of mixed stores to the various ships on our errand-boy round of the dockside. By the start of the second week, I had become quite used to the mad antics of Garry, as our little toy train steamed about its duties in a very erratic manner. Short, sharp spurts of speed, if Lady Luck had smiled on him, or a slow crawl, if she had turned her back on him.

A winning streak would also arouse Garry's strange sense of humour. We had called daily on a BP tanker, whose deck crew were Asian. The toilet routine of their head man had not gone unnoticed by Garry, as he watched the man go between two wagons to perform his bodily functions each day.

'Watch this,' he said, as he quietly took the long shafted fire-dropping shovel from the tank top. His eyes sparkled with mischief as he slid the shovel across the grassy surface of the four-foot to rest beneath the Asian's bottom, who was crouching the other side of the wagon. It seemed to be an eternity that the shunters and myself suppressed our laughter, while Garry waited patiently for the offering before withdrawing his shovel. To this day I can't recall a more unexpected situation, as the poor Asian stood up, fastened his trousers to his braces, then turned around to look down behind. His face portrayed utter disbelief, whilst his actions became erratic, as he began frantically to search the area on all fours. Finally he slouched up the gangplank to his ship, stopping now and again to look back on the scene of his illusion, whilst Garry, free of any suspicion from the pathetic little man, gave forth with a burst of tick-tack.

When I asked him what was the message, he gave one of his rare chuckles and said: 'The phantom strikes again!'

During the third week shipping became extremely quiet, and it became difficult to warrant our existence. On a few occasions we puffed up and down the dockside with two empty vans for no other reason than to appease the yard foreman, who said we set a bad example of malingering to the other shunting engine crews. Eventually we found a nice long empty siding where we could form a card school, secure from all intruders, as we nestled against the stop-block. I watered down the fire to keep her quiet – a punishable offence if ever there was one – and we settled down to the true meaning of 'Light Duties'!

After countless games, the head shunter looked at his watch. 'That's it then, boys. Time to finish. Let's head for the shed.' His face sallowed suddenly, as he looked up the siding. 'Oh no! I'm blowed if I heard that ruddy lot come in here.'

Just 10 feet away from our buffers was a half-full train of coal. I looked up at our steam gauge: 60 lb pressure. Just enough to take us on the shed. But worse was to follow, as I opened up the firebox doors. The fire was out. Even with full pressure, the gallant little 'Coffee Pot' would never tackle twenty 20-ton wagons of coal, but with no fire and 60 lb on the clock, the disgrace was too much to comprehend, as we remained helpless in our prison.

The awful truth dawned when we learned that the yard pilot had gone elsewhere for a few hours. To a constant tirade of abuse from the yard foreman, we sat, heads bowed in shame, whilst waiting for help. It arrived at long last in the form of a 5600 class tanky, which had brought a train of mixed goods from another yard. We suffered the degrading shame of being the forlorn little tail-end Charlie, as the '56' waltzed us towards the coal hoists of the High-Level. But our disgrace was not to end there, for no one would risk the conspiracy of taking us on the shed, as we waited hopefully for some kind loco crew to heed our plea. By this time poor little 784 was completely cold and we were three hours overdue. The picture must have portrayed a mixture of pathetic humour – four men marooned on the footplate of a tiny steamless loco, trying to hitch a tow onto the shed, which was virtually a stone's throw away.

Suddenly came the unmistakable hiss of a Peacock steam brake exhaust. Only one loco, other than 784, had such a brake – 783. She drew alongside her cold and seemingly shivering sister with the warm glow of an iron fairy godmother, as the two drivers faced each other from their cabs. No tick-tack came from the subdued Garry, as our shunters explained our plight to the other light duty engine crew during the fading light of that evening. They had been on legitimate overtime, hauling the mess-coach of the permanent way relay gang.

Twenty minutes later, under cover of full darkness, the coupled twins arrived over the drop pit to the unwelcome approach of a rather nasty firedropper. He had completed ten years of firing when he developed colour blindness. He had the option of taking up locomotive shed duties, such as berthing, or shunting, but, as a protest to no one in particular over his misfortune, he decided to sever all connections with driving and firing to take up firedropping. Consequently he became bitter towards footplatemen, taking a delight in performing any injustice possible on them. His face broke into a cynical leer, as he swung up to our footplate. 'Strange to see you pair of blighters come on the shed together – and so late!'

The effect of the traumatic incident was still with Garry, who, minus any tick-tack, brushed past the firedropper.

The following morning during engine preparation, foreman Edwards appeared, speaking across the top of the motion to Garry as he oiled the shaft.

'Don't think we're going to pay you for that bit of overtime last night, Garry. More than likely we'll take a bit off you.' He gave a broad smile as he tilted his bowler. 'You can hardly be expected to be paid when you're running round with no fire and no steam, so to speak. Mind you, the firedropper was pleased! He said he wished they'd all get towed in with no fire in 'em.'

It was midday before Garry spoke. 'Nip up to the fishmongers, mate, and get us a pair of kippers.'

When I returned I thought he was going to enjoy a succulent shovel feast, but the pair of herrings went straight into his 'tommy tin'.

On the Monday morning of my last week with Garry, I noticed that despite the bitter cold weather, the firedroppers sat outside their mess-room. Garry was full of smiles and tick-tack as he shouted across to the firedropper – or Judas the Clinker, as he called him – 'Too hot inside for you, mate?' The cold-looking man scowled back, for it was no secret that a terrible smell existed in their cabin, despite the efforts of fumigators and disinfectants.

As it neared midday, once again Garry asked me to buy him a pair of kippers – 'And this time I'll eat 'em,' he said, as he handed me the money.

'Well, what did you do with the others?' I asked.

He tapped each elbow in turn, then held his nose in tick-tack fashion. 'Nailed 'em underneath the firedropper's

the handbrake should have been on. As I quickly applied the train brake, I noticed Dai, impervious to the situation, admiring the station flowerbeds. There could have been a serious runaway if I also had left the footplate, and the fact was that we were both to blame, for I had not applied the handbrake.

Even more ludicrous was the fact that, Dai, quite unperturbed by his near accident, left the ejector on every day of the week, but I had learned my lesson, and the handbrake was engaged in readiness. When I spoke to his regular fireman the following week, he passed the matter off with a laugh and a shrug, saying: 'You have to treat Dai like a child.'

Barry Shed also suffered a steady flow of 'made up' drivers from mainline sheds, who had no knowledge of the South Wales coal train industry. They were excellent enginemen on fast vacuum brake trains, but the toy trains of the valleys, as they called them, were a different world. There were no hard-and-fast rules to bring a coal train down to the docks. At each gradient the driver would stop, then pull away, while the brakesman started to pin down brakes. When the driver felt that he could control the train, he blew his whistle to tell the brakesman. From there on the fireman would use the handbrake, and the vacuum brake was the last resort. It required years of skill on the part of the driver to acquire the feel of the braked wagons, and the experience usually stemmed from his fireman's days.

I remember working a night coal train with a newly appointed driver to Barry, a cocky, braggish type of individual from Old Oak Common Shed. As we struggled to head our train of empties to the colliery with a '56', and despite the fact that those fine locos were designed for the South Wales work, he constantly let forth with a flow of sarcasm. 'Ruddy Hornby train set work this is, mate! Wish I was back on the "Runners" instead of this snail crawl.'

I was too busy keeping steam pressure up to take much notice of him, but I knew that on the return journey, when my work would be negligible, his driving expertise would be stretched to its limits.

About 2 a.m. we left the colliery with forty 20-tonners of Welsh large for Barry Docks. The early moon was rapidly losing its brillance, as we stopped at the first gradient, a 1 in 30 slope of six miles. The brakesman could barely have secured three or four brakes when my driver blew the whistle. This was the fatal mistake for any newcomer to coal train working, for, when starting away, most of the train is on the flat, but once all the train is on the gradient, the engine receives a massive push, as if indeed it were just a toy train. As we began to gain speed, I slowly turned the hand-brake. 'No need for that, mate,' said my driver, and he began to sing loudly, but in a rather nervous manner. Suddenly we received our push from the wagons behind. He jumped from his seat and applied the full force of the vacuum brake, causing the wheels to skid as they became locked by the brake blocks. At a speed of between 20 to 30 mph, we were obviously out of control, and I began to prepare to jump for it, if we passed any signals at danger. The embankment took on a bonfire night effect, as myriads of sparks played from the friction of wheels locked against rail. Luckily all our signals were green, and with a sigh of relief, we came to a standstill on the flat some six miles further on. When my driver blew off his brake, we proceeded with a strange motion, and an equally strange noise – flat tyres, worn down through the long skid!

My driver went sick the following night for quite a number of weeks, after which he visited Swindon to see the Company doctor. He returned to take up dock shunter work. The rumour was that he suffered a nervous breakdown after the trauma of that night, and he couldn't face valley work again.

I remember some older drivers, whose names are still legend with today's railwaymen. One of them, Dick Germon, was a spritely little man, a good engineman, an inveterate gambler, and an equally compulsive practical joker. One story relates how he had a relief fireman to work a valley coal train. Wenvoe Tunnel was the scene for Dick's joke. As the '56' entered with her train of empties at a good turn of speed, the fireman continued to build his fire after the ravages of the first incline. Suddenly he let out a scream of terror in the half-light of the smoke and steam of the footplate, for protruding from the bunker hatch was a blood-stained human hand! It originally belonged to a tailor's dummy but Dick had found it on a refuse dump and added a little red paint to complete its macabre appearance.

Some time later, in the same tunnel, Dick kept his eye on another relief fireman, who was on guard against any of his jokes. Eventually he told him to keep a look-out ahead, and, as the fireman peered into the dark through the eyeglass, a grotesque green illuminated face appeared from outside. Once again Dick's victim uttered a cry of alarm, but became even more terrified when he discovered that there was no driver on the footplate. The fireman became frantic as the train steamed on, shouting helplessly for Dick, who suddenly appeared on the fireman's side of the footplate. When he told the fireman to look for the distant signal, he quietly took the gauge glass lamp, which could double as a signal lamp, and, climbing out of the cab and over the boiler, contorted his features with the use of his false teeth, as he held the green light of the lamp to his face.

His most daring joke, however, was performed for all to see, and it was a mystery how he evaded the strict discipline of the pre-war Great Western. It was during the time that he was a passenger fireman. The day was Derby Day, and, true to his gambling instincts, he travelled the mile or so from Barry Island to Barry station astride of the boiler, with his hat turned back and using a piece of rope around the smoke stack as reins.

Practical joking, however, went part and parcel with the job, and railwaymen would undertake immense preparation for their trickery.

Ex-Barry Railway tanky number 268 had the nickname of 'Killer' – three men had died under her wheels over the years. About five o'clock one morning she made her way slowly along the dock road toward the shed, after completing three shifts of shunting duties at Rank's flour mills. Frank Poulton was the driver – a good engineman, but constantly on the lookout for a practical joke.

Frank suddenly applied the steam brake. 'Did you feel that, mate?' His sleepy fireman shook his head. 'Well, I'm sure we went over something,' said Frank. 'Take a look your side.'

The yawning fireman slowly descended his steps, then suddenly reappeared on the footplate, white faced and breathless. 'There's . . . there's a ruddy head down there, mate. Gone right over him, we have. Hell, we was only doing walking speed! This blighter's at it again. She's a ruddy murderer, I tell you. She can't stop killing.'

Frank Poulton fought to suppress his laughter. 'He's hell of a length, mate! His feet are sticking out this side. Must be all of seven feet long.'

The fireman peered cautiously over the driver's side. 'Looks like suicide to me, the way he's laid himself across the line. What a way to go! He could have done better than that – like a passenger train or fast goods! It would have been a lot quicker for him.'

'Well, you'll have to go and tell the dock police,' said Frank.

The fireman leaped straight from the footplate into the next four foot, without taking a second glance at the corpse, and ran along the dock road towards the nearest police lodge.

Within a few minutes, 268 caught up with him, as he ran breathlessly on. 'It's all right, mate! Don't bother with the police. He's come around a bit. He's up here on the footplate with me,' said Frank.

'What the hell do you mean?' said the fireman. 'He's in two.'

'Well, come an' see for yourself,' said Frank.

The fireman gingerly climbed aboard to face the divided body of a tailor's dummy.

Frank Poulton explained later to a mess-room audience how he had arranged with the crew of an engine that was going to the shed earlier to lay out his dummy in suicide order.

My own driver, Bill Bouchier, was also a character. He was an excellent engineman, who believed that firemen should spend at least a quarter of the shift in driving. This, of course, was quite logical, for how else was he capable of handling a train when the day of his promotion arrived?

The period was soon after nationalisation, and British Rail were introducing several new innovations. Some were quite sensible, and some were downright stupid.

For as long as anyone could remember, the shunting pilots of Barry Docks carried a red light on the top brackets at each end of the loco. They could be seen above all the rolling stock from a great distance, and collisions were extremely rare. The new BR plan was to use four lamps – two red and two white – on the bottom brackets, but completely hidden in the endless sidings of rolling stock where they worked.

We had just arrived at Cadoxton sidings with a coal train from the valleys, and Bill Bouchier was anxious to attend a committee meeting at the Railway Club. 'Put some water in the bucket, mate,' he said. 'I'll have a wash while you take her on the shed. Don't bother about the mainline, it's too busy tonight. Go the docks way.'

As I drove the loco along the docks down-line, forgetful of the new pilot lighting system, my eyes scanned the area for single red lights. Suddenly, in the inky blackness of the dockside, a train of coal empties loomed before me broadside on. The pilot engine which hauled them carried four lights, but I could see none. We cut the train in two, derailing our own engine just as my driver was washing his face. The impact threw him forward, and even with the excitement of the moment, I laughed at the spectacle of him lying on the footplate with the bucket on his head.

Bill didn't get to his committee meeting. In fact, the Club was well and truly closed before the breakdown gang had put us back on the rails. He told the accident inspector at a later meeting that he was driving at the time and, due to the new ineffective lighting system, he would accept no blame for the accident. He was cautioned, but a few weeks later all dock pilots reverted to the old lighting system.

I remember during the latter part of the war working the afternoon shift of B15, which could be a bit of a grind. We relieved the crew at Barry station after the engine had completed her Vale of Glamorgan 'maid of all work' duty. The engine was ex-Barry 259, the old tanky with her original badly worn boiler. There couldn't have been a worse engine for the job. It was a foregone conclusion that we would take twelve hours to complete the round trip to Newtown Yard, Cardiff, a journey of less than forty miles.

The reason, of course, was the war. It appeared that we were the least essential goods train in the country, and a good 80 per cent of our time was spent hogging the loops whilst train after train steamed past. This was soul-destroying enough as your head moved back and forth to the board like a Wimbledon spectator, but with 259 it was torture. She just wouldn't keep quiet. She had to have a decent fire to haul away a train of forty mixed when the call came, but with both dampers closed and a black fire she'd still scream away from the safety valves.

I recall the words of Joe Waite, my driver, during a peaceful moment when a gently singing injector soothed the hoarse vocal chords of 259's safety valves. 'You know, mate, we've got to be compatible to some degree – I spend more hours with my fireman than I do with my missus. Lord knows it would be ruddy hell if we didn't get on.'

Joe was the salt of the earth, a kind-hearted jovial character, and an excellent engineman. His remarks reminded me how important was the compatibility of fireman and driver, two men confined on a small footplate for an average of ten hours per shift in often comfortless and stressful conditions. Although most crews integrated well, there was, of course, the odd tyrant on both sides, when the day of a general link change could not arrive soon enough for the long-suffering victim. That day was like a lottery result. Crews signing on and off would rush to the new roster case, and conversation would flow. 'Who's this new bloke I got, mate? Can't say as I know him.' Or, 'Now you'll have to work, mate, he thinks the lever's only for going backwards and forwards.'

George Lawdy was the driver who commanded my highest respect, a mainline goods man who had the misfortune to suffer me on my return from Westbury to Barry. I was seventeen and a half and still as green as grass when I signed on at 1.35 p.m. for the spare link. The time clerk handed me a set of engine keys for 5619. 'You're working B30. The Treharris regular fireman has had to work a passenger.'

Many drivers would question a very young fireman regarding his past record, often refusing to take a greenhorn on a mainline trip, but George Lawdy, from the outset, treated me like his regular mate. At the time I wished he would question my firing background, for I had never worked a valley coal train before. The engine, like most of her class was perfect, but I had misgivings when I noticed her bunker was only half full of best Welsh large, for what might we get at the coal stage!

During my cleaning days I had helped many a mainliner fireman to prepare an engine and I put this to good use. One of the most important points was sand. If there was the slightest hint of any wet sand in the boxes, the fitter and his mates had to be called. It's a strange phenomenon how water splashing off the wheels against the sand pipe could travel right up to the box and contaminate all the contents. On many occasions a tap on the pipe with the coalpick would unblock the system, but on that day I took no chances. The front pipes were only just trickling and I went to the fitter's cabin. I was giving myself a lot of extra work when I asked the two men to empty the boxes, but the stories of sandless engines in those valley jobs haunted me. I filled up with red-hot sand from the drying house, and felt more secure.

With an all-out effort to appear as a veteran fireman I found time to fill the cylinder lubricator for George, and with a fatherly smile from him we proceeded to the coal stage. I found it difficult to suppress a nagging desire to shout up 'It's my first valley trip, best large please,' but the sound on the cab top soon told me my prayers had been answered. George came out with one of his rare pieces of advice when the coaler shouted down that we were finished. 'Nip up and trim it, son, and ask them to put another one on.' We picked up our guard near the drop pit and collected our brake van for Cadoxton sidings.

The train was light, forty empties for Treharris. As usual, and against the Rule Book, the guard gave us right away, and walked back to his van. We'd never see him wave from the rear of the train in those curved sidings.

The '56' seemed to laugh at her train, whisking it up Wenvoe bank with ease. I felt that I'd overdone it on the sand routine, but I was unaware until a few days later that a

Cadoxton pilot had given us a liberal shove up the normally difficult incline.

It was dark by the time we arrived at Treharris, and George reminded me to pull down the wartime blackout shutters, because trains could be a prime target for the enemy. I felt pleased with myself, for there had been no set-backs, and not once had George interfered, or even advised me, although I was aware that he watched my every move.

I had fallen in love with 5619 as we sat in the cosy atmosphere of her enclosed cab that night in the colliery yard. My bread and cheese sandwiches and my mug of strong tea were a meal fit for a king as I chatted to George. I had completed the worst part of the job, now it was downhill all the way home. Twelve hours later after hauling fifty loaded wagons of coal back to Cadoxton through the constant threat of an air raid alert, we signed off.

As we walked past 5619 at the end of the drop pit on that full moon night she appeared as quiet as a lamb; it was hard to believe that she had barked her way up the valleys at full regulator. The next day, the next week, and the next year, and for many years hence she would continue to pit herself against those tortuous gradients.

I lived quite near to George, and as we neared our houses he said completely out of the blue, 'I'm cleaning out the shed, do you want my motor bike? Come round tomorrow and pick it up.' A 1930 BSA 250 cc. George had ridden it only once from new, when he fell off it, and decided that motor cycles were not suited for him.

In my young mind it looked like something out of the Ark, but it was virtually brand new. I could have got a petrol ration for it, but I was ashamed to take it on the road. It was a constant source of embarrassment when George asked me how I was getting on with the bike, when it remained locked up in my shed.

A year or so later George, who was by then a passenger train driver, suffered a fatal heart attack as his train approached Barry station. His fireman brought the train into the platform and there were no safety problems.

For many years I kept that motor bike, subconsciously as a form of a shrine, sitting with it in my shed and recapturing that wartime journey on 5619. George knew how to handle engines and firemen.

The greatest engineman of Barry Shed must have been Driver Dunn. Reluctantly retired by the GWR in 1937, his love of steam grew stronger with its absence. A very fit and agile man with a smooth tongue to match, he joined the crew of the collier S.S. *Ferranti* sailing from Barry dock. It was no mean feat for a man of his age to fire a ship's boiler sometimes in atrocious weather. To this day, he is a legend among the modern enginemen of South Wales. The son of one of the Barry Railway's first drivers, Bill Dunn worked at the age of ten with his father. 'It was a kick up the backside or a clout across the ear, if you dared to fire her wrongly,' he'd say to his own firemen, as he recalled the past.

His record as an engineman was unblemished. Never late. Never sick. No engine failures. An extremely healthy and determined man, devoted to the engines of the Company. He would trust no fitter, and often spent his Sundays carrying out maintenance work for no pay. He claimed to be the great grandson of George Stephenson, referring to him as the 'Messiah of steam'.

'Engines are human,' he'd say. 'Treat 'em right and they'll not forget you.' He wouldn't dare grace the footplate of a diesel during one of his retirement visits to the shed. 'Where do you wind 'em up?' was his sarcastic remark.

It was on such shed visits that he would observe all the activities with a critical eye. I recall how his walking-stick rapped the door of a '56' that I was preparing. 'You're making smoke, lad, and your injector's running dry. Mend your ways or you'll never make an engineman.'

His sea-going career continued into the Second World War, by which time his knowledge of engines had earned him promotion to third engineer. No one aboard the *Ferranti* was aware that he had trimmed his true age by a good ten years when he joined her. But the war was taking its toll, and Bill was no swimmer. To a reluctant chief engineer he bade farewell, as he left the *Ferranti* to work once more on the footplate.

For the next twelve years he drove and fired the factory locomotive of the Aberthaw cement works. During that time no works' fitter was ever allowed to touch the immaculate saddle tank, and any Great Western loco which entered the works could be a victim of his tirades, as he compared the old days. 'Had my own engine then,' he'd say. 'Used to give my cleaner a halfcrown a week to keep her like new. Thought more of her than my old woman!'

Progress, however, was Bill's downfall. As more and more lorries used the works, the company's insurers queried the age of their loco driver. The management reluctantly retired Bill yet again, but the remarkably agile pensioner was far from being ready for the shelf. It is difficult to record his true age at the time, but somewhere around seventy-five, he was successful in securing yet another job as an engine driver! A local chemical company had recently begun operations at Barry, and George William Stephenson Dunn was the obvious choice to drive their loco. For a further five years Bill drove his little loco around the confines of the factory complex. His days were happy, with the feel of the regulator, the smell of the steam, and the roll of the footplate. But the high-railed enclosure of the factory spelt prison to him. His eyes would mist over with visions of the past whenever a British Rail mainliner called to pick up a train.

He eventually retired for the last time at the age of eighty. 'Failing eyesight,' said the Company doctor, as a cover-up for dispensing with a steam fanatic, who would just not grow old.

Yet there was still twelve years of active life left in this boilerhouse of a man. Having mixed affections for ships and locomotives, he divided his attentions between visits to Barry Shed and the docks' entrance. But his idols still had to be powered by steam. In his eyes no other form of power existed other than steam, be it a small stationary engine, or the mighty triple expansion unit of a ship.

If there are such places as heaven and hell, I doubt that Bill Dunn would have been too concerned if he had gone to the latter, as long as the fires generated steam.

1 The author as a young fireman mimicking an Edwardian pose between his mother and sister, 1945

2 Barry Docks, the world's greatest coal-exporting port, *c.* 1938. Number One Dock (*top right*) is served by the coal-hoists of High Level and Low Level; Number Two Dock (*bottom right*) is served by East Dock hoists

3 First of the American 2-8-0 locomotives being unloaded at Cardiff Docks, South Wales, December 1942. These were the equivalent of liberty ships of the war-time railways

4 American 2-8-0 locomotives stored at Tonteg on the Barry–Rhondda line in readiness for D-Day

5 1683, the 'Big Yank', at Barry Shed, 1948. She was laid up for several months with shipment damage mainly to her frame and ashpan

6 The firedropper starts on ex-TVR 0-6-2T 361 at Barry Shed. This was the first engine to be painted in British Railways livery at this depot, *c.* 1948

7 5600 class 5634 heads a long holiday excursion across the causeway into Barry Island. Note the waving handkerchiefs as passengers near their Mecca during the summer of 1952

8 'Castle' class 5052 *Earl of Radnor*—the 'soup spiller'—at Paddington in 1952 after heading a Cardiff express with (*l–r*) Guard Bill Hooking, Driver Bert ('Dogger') Jones and Fireman Mark Macleod

9 Ex-TVR 0-6-2T 312 starts to blow off in readiness for her gruelling trip 'down the Vale' from Barry to Bridgend, June 1953. These locomotives were fickle and the journey was known as the 'switchback'

10 The rear of Barry Shed, 1954: ex-Taffies and GWR panniers appear to predominate the scene

11 Roof repairs enabled this photo to be taken inside Barry Shed in 1954 when the cleaners were working on BR 2-6-2T Class 3 82000 series 82042

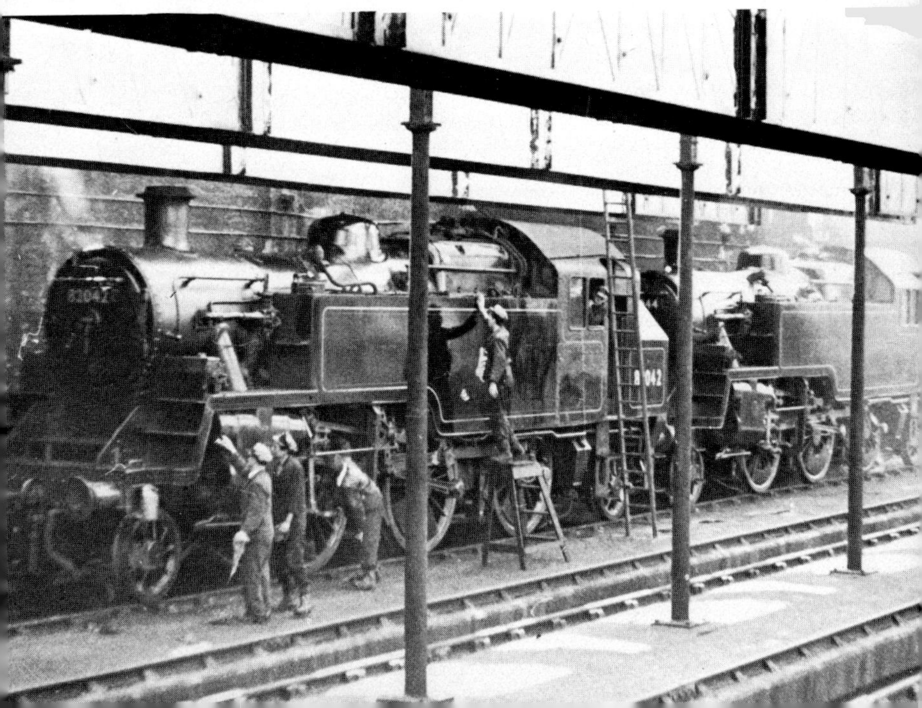

12 The end of a hard shift as 5600 class 0-6-2T 6614 returns down Wenvoe Bank intc Cadoxton Yard with a train of coal, *c.* 1954

13 7022 *Hereford Castle* heading the 10.00 a.m. Red Dragon express from Cardif General station to Paddington, October 1956. She was broken up at Bridgend ir 1963 after nearly 750,000 miles to her credit

14 A rare photo of 7012 *Barry Castle* being diverted through Barry during mainline repairs on a Sunday in 1959. She was broken up at Cashmores Great Bridge in 1964

15 4080 *Powderham Castle* picking up her paces as she leaves Cardiff heading the 10.00 a.m. Red Dragon express for Paddington, August 1960. She ended her days at a Newport scrapyard after completing nearly two million miles

16 A busy morning at Cardiff (Canton) Shed, 1958: 'Britannia' 70029 *Shooting Star* (*l*) is being prepared for the 1.13 p.m. Salop; note the immaculate appearance of 'modified Hall' 6999 *Capel Dewi Hall* (*r*)—ready to work the 1.00 p.m. Brighton partly over SR lines and rivalry is strong

17 8 September 1962: a sad day for Cardiff (Canton) Shed as only four locomotives are seen on its last day of operation before closure

18 'The toughest little worker of the GWR': 5700 class pannier tank 8728 heads the Cardiff goods into Barry in 1963. This was the class the Americans named the 'matchbox'

9 South Wales Valleys' workhorse 0-6-2T 5692 with her van at Biglis Junction, Barry; she was one of the last steam engines to work in the Barry area during 1965

20 Decay among the buddleia: a pair of once-proud GWR 'Halls' stand bereave amongst their funeral flowers at Woodham's Yard, Barry

8 · Firemen

Not unlike the drivers, firemen were a mixed breed of men. There were those who lived for the job, taking a great pride in their skill of steam keeping, whilst their footplates were as the proverbial palace. Their personal appearance was also an eternal mystery to me: well-polished boots, crisp, clean overalls, complete with collar and tie!

Despite my efforts to join their élite ranks, I always appeared dirtier than a coalman, even after constantly hosing water over the coal. Unfortunately, I was the type the shed foreman pounced on eagerly for such menial work as ashpan cleaning and coal stage duties. Eventually these tasks were forbidden after the only strike threat that I could recall, but our objections were well founded when the Rule Book stated that our appearance should be of high standard at all times. One session on the coal stage, or the drop pit, would put paid to a set of clean overalls.

Tobin, my old friend of cleaning days, continued to stumble his way through a railway career beset with hilarious situations during his fireman's days. For example, he was on spare duty at Barry Shed and, during his break period, he was just about to enjoy a pair of kippers, when the foreman rushed into the mess-room. His face fell when he realised his only choice was Tobin, but he was desperate. 'Quick lad! Over to number three platform and relieve the fireman on the down Vale passenger. He thinks he's got appendicitis.'

A little later, the driver of that train threw up his hands in horror, as Tobin ambled along the platform, picking at his kippers wrapped in a piece of newspaper.

He was not the best of firemen, and, somehow or other, well-renowned good steaming engines failed to respond to his attentions. But the particular loco that Tobin approached that day was a rather sensitive engine: number 381, an ex-Taff Vale Railway 0–6–2 tanky, which was extremely patriotic through her love of best Welsh steam coal. Anything less, and she took on changeable moods, and a good fireman was needed to handle her. That day her bunker contained imported American coal, quite good fuel really, but not suitable to the fussy appetite of 381.

The Vale line between Barry and Bridgend was known as the 'switchback', 30 miles or so of undulating inclines, bridges, viaducts, and severe curves. Poor Tobin was in trouble after the first few miles and, from there on, the steam pressure gauge rarely went beyond 150 of its 200 lb maximum pressure, and the water-level constantly played hide and seek in the bottom nut of the gauge glass.

At Bridgend, despite a bitterly cold day, the coach heating steam had to be turned off for the return journey through to Cardiff, as the driver fought desperately to coax a full head of steam from the fickle 381.

He was unaware, however, as he prayed into the firebox, that the Bridgend station-master was escorting a very senior member of British Railways management towards a first-class compartment on his train. Thirty minutes later, as 381 gasped to a welcome stop at one of the many Vale halts, the great man of British Railways came along to the footplate to face a very sweaty, dirty, miserable Tobin. 'You've no heat in the coaches, fireman!'

Even the driver, whose patience was rapidly eroding, believed him to be a railway eccentric, the type who knew the job better than the engineman – at least on paper.

Tobin broke into one of his doleful grins. 'I know that, mate, and you got two chances. We can all stay here and keep nice and warm, or you might reach Cardiff frozen stiff, but you can't have both!'

The guard's whistle was blowing repeatedly for 'Right Away', but completely ignored by Tobin, his driver, and their British Railways superior.

'I tell you what, mate,' said Tobin, 'I'll block off the steam on the first coach and I'll put a few pounds into her. That should stop you freezing to death if you care to ride next to the engine.'

The cold passenger turned on his heel. 'Oh, don't bother! For goodness sake get this ice-box to Cardiff. There must be a warmer train that I can change to somewhere on this system.'

Tobin shouted after him as he stamped off along the platform, 'You can have a loan of my overcoat if you want to.'

They did eventually reach Cardiff to disembark a very cold-looking train of passengers – including the British Railways manager, who presented his card to the pair of stunned enginemen. But he had accepted Tobin, and his humour in good heart, and nothing more was heard of the incident.

Tobin, however, was like a magnet to misfortune, often quite unwarranted, despite the fact that he was a poor engineman.

The long haul out of Cadoxton Yard with a train of coal empties for the South Wales Valleys went on for about six miles. Yard foremen seemed to take a sadistic satisfaction in searching high and low for possibly one single wagon to make up a full load. Quite often the progress of the train was slower than walking speed.

Tobin was fireman of one such train on a warm July afternoon. He had a good '56' class loco, which responded well to even the worst fireman, but a massive boulder of coal – possibly over two hundredweight – blocked the bunker entrance as the train began to climb.

Half-way up the gradient, at the foot of a steep embankment, lived Widow Fouracre in the railway house of her deceased platelayer husband. Most firemen would throw down a few lumps of coal as they passed, and Tobin was no exception as he struggled to free his bunker from its huge obstruction. The lump, if ever she could break it up, would warm her house for two weeks at least, thought the good-natured Tobin.

In a bath of perspiration, Tobin eventually manhandled the great black lump to the engine door, just as they approached the cottage. The marksmanship was perfect – too perfect. Down it rolled in a singular avalanche, through the neat garden gate, and completely through the little black front door, taking it clean off the hinges! Fortunately, Widow Fouracre was out, but she must have been quite shocked on her return to be confronted by a two hundredweight lump of coal on her hearth rug!

The return journey proved just as traumatic for another individual, when a stray sheep ran in front of the engine, just before it came to stop at a signal. Dusk was closing in fast when Tobin and his driver struggled to haul the carcass aboard the footplate, for wartime ration books had not long been dispensed with, and such a good supply of fresh mutton could not be left for fox food. Tobin was pressed into being the butcher with nothing more than the driver's pocket knife and a small wood saw, which every loco seemed to carry for firewood preparation. By the time they had reached their terminal of Cadoxton Yard, the beast was cut into several very crude joints, whilst Tobin was sodden in blood from head to foot.

It was quite dark by the time the guard had walked from his van to join the engine for the journey to the shed. As he climbed the loco steps coming face to face with the grotesque figure of Tobin, illuminated by the flickering fire, he gave a scream of terror, and ran back up the siding.

It took over an hour to find the guard, who was convinced that Tobin had gone mad and had savagely murdered the driver, but later he eagerly accepted a leg of mutton.

To return to the 'pristine' firemen – they were an amazing breed of men in my eyes, and they were most certainly stereotyped. Their mode of dress was fashioned on the firemen of the 'runners'. Overall jackets were tucked into the trouser at the waist, and braces were worn outside. Cycle clips, elastic bands or socks secured the trouser bottoms to stop the dust going up their legs, and the head-dress was a tammy. This style of dress seemed to go hand in hand with the 'Castles', but I suspect there was a strong status symbol in the trend, for many a fireman of the humble pilot dressed that way.

The tidy firemen also attended the Mutual Improvement Classes on a Sunday morning – what lucky employers the railways were in those days!

And on the subject of devotion to duty, it was an unwritten rule to sign on at least 15 minutes before time. Of course that was a necessity on many occasions, since 45 minutes' preparation time could fly. There were several trips to the stores for lamps, coal picks, shovel, prickers, bars, and if the engine tool box was unlocked, a bucket of tools.

The first duty on reaching the footplate was to check the boiler water level, which was a religion that took preference over all else; the fear of dropping a fusible lead plug was instilled into our very soul from the first day of our railway career.

I remember one of my very early firing turns, when after several turns of the blow-down cock on the gauge frame, I could see no water. I rushed across to the foreman's office to confront my old friend Mr Chidgy. '271, she's got an empty boiler,' I said in a superior manner. In his usual way he got up from his chair to walk past me as if I didn't exist. I followed him across to the shed and my engine, where he half closed the gauge frame lever, and opened the blow-down cock. A water level appeared in the glass. He climbed down to the shed floor before he spoke: 'Trouble is, you got too much water, mister, she's not a steam engine, she's a bloody hydraulic engine.'

I knew the cause well enough – the steam-raiser. They had a couple of dozen engines to look after and quite often they would forget an injector they had left working which over-filled the boiler. Some firemen would do the whole of the preparation job frequently on pilots, leaving the driver to enjoy a good chat session in the mess-room, who claimed it was the best way for the lad to learn the job.

Many firemen had part-time occupations elsewhere, such as window cleaning and house painting; others had small-holdings of poultry, sheep, pigs or goats. As for me, my interest in small boats kept me quite busy with no revenue such as my colleagues enjoyed. Of course, it was the wide range of shifts that allowed us to follow our various pursuits.

I recall a particular week of night work during the month of June. I was keen to get my boat into the water, but she needed a lot of work, and the weather was glorious. I was on a Cadoxton pilot, a busy job despite its title. The first night passed quite well as far as my stamina was concerned with the exception, of course, of the 'Death Watch', that is a period around four in the morning when the body is said to be at its lowest ebb. I believe most railwaymen can vouch for the wisdom of this belief, for we would certainly begin to flag about that time, only to revive considerably when dawn broke.

As I walked home after my first night shift listening to a beautiful dawn chorus of birdlife to herald the promise of a perfect June day, I decided to miss bed, and spend a day on my boat. I began to feel the effect of no sleep during the next day's shift but strange to say, as soon as the 'Death Watch' passed, and dawn broke, I came around to spend another day on my boat, despite my self-promise of a good day in bed.

By the time Friday evening arrived, my only sleep for the week had been a few cat-naps, and I was shattered. I remember putting the pricker into the firebox when preparing on the shed, then beautiful oblivion, until my driver woke me up. I was still holding the fire iron, but it had become red hot and bent under my weight, and I was kneeling on the footplate. I don't know how I survived that night – my boat was looking perfectly shipshape whilst I was an utter wreck.

During the night as we hauled a train of coal from Cadoxton Yard to High Level, I lit a cigarette. The engine was going well with a full firebox and a healthy boiler, and I required very little of the rhythm of her steady beat to lull me into deep sleep, as I leaned against the tank. Suddenly I became aware of an acrid smell in my nostrils, and a burning in my throat as Bill Edwards, my driver, and the two shunters made over-zealous efforts to smother my smouldering overalls.

After a week's work they had absorbed a fair amount of paraffin and oil and the cigarette had ignited them. The full extent of the damage was more embarrassing than anything else, for the course of the burn had travelled neatly around my thigh causing the complete leg of my trouser to fall off.

I had to suffer the wisecracks of my three colleagues for the remainder of the shift, as I continued my duties with the appearance of a circus clown, but I could not face a couple of dozen men on the shed. As we approached the drop pit I asked my long-suffering driver to sign me off, and as the dawn began to break, I bade farewell.

I don't know if one bare leg constituted indecent exposure

in those days, but I received a strange look from the town policeman as I rushed past him on the way home.

I remember letting myself into the house, taking off my boots, and one-legged overalls, before going to the bathroom, but on the way I had a compulsion to look at my bed. Words could never describe the sight of that emblem of rest, that contraption of peace. Dressed in underpants, vest and about a pound of coal dust I was drawn to it like a magnet. Eighteen hours later I staggered down the stairs to confront my mother in the living room. She gave me that look that only mothers can give. 'You've missed your turn, son. I tried to wake you, but it was impossible, you were like a corpse.'

We both broke into spontaneous laughter at my dishevelled, dirty appearance, and by some magic that only mothers seem to possess, a steaming hot meal was produced. 'There now, son, get that down you while I change the bedclothes – then you can take a bath and carry on sleeping. You and that old boat, I knew something like this would happen.'

On the subject of beds, however, many enginemen will recall with little nostalgia the dreaded 'double home' work. This could be goods or passenger work where the outward journey lasted over the eight-hour day, and generally speaking this usually occurred on the Swansea and Cardiff runs to Old Oak Common, or thereabouts.

The GWR 'approved' lodgings in the Paddington area were little better than doss-houses at about five shillings for a bed and meal. The amount even in those days seemed trivial, but each average size bedroom contained at least half a dozen single beds, which were rarely empty around the clock, and many was the time that engine crews had to wait for their beds to be vacated. They had a welcome warmth on a cold winter's day, but the system was not exactly hygienic.

The unmarried young firemen of those days were a wild bunch, and with their quota of free passes and privilege tickets, a day from Wales to London was quite the norm. The 8.00 a.m. from Cardiff to arrive at Paddington at 10.50 a.m. was the usual train. A day sightseeing followed by an evening at the Hammersmith Palais passed like wildfire, more so if we succeeded in meeting up with a girl at the dance.

Many a young fireman has reached Paddington, breathless and lipstick smeared, to see the tail lamp of the last train as it snaked out of the station. This was the slow newspaper train stopping at every blade of grass to arrive at Cardiff General at 5.35 a.m. but as the first Barry train left at 5.40 a.m. it was just right, and we had a good sleep in the bargain. Those who missed it could always find a coach in Paddington sidings, but it was wise to check with the shunters first. One fireman failed to do so and ended up in Newton Abbot.

The Railway Club, branded as the 'Snivellers Club' as a long-standing insult to the founder members for being Company men, was the focal-point for all enginemen to meet. It consisted of a corrugated iron-clad building by the railway embankment close to Barry locomotive shed. Smartly dressed, off-duty firemen would discuss all aspects of their lives with their begrimed colleagues, who had called in at this oasis to wet the coal dust. It was not unusual for the steward or one of his helpers to serve thirsty firemen with a pint of beer to be told, as he returned from the till with his change, 'Refill, please.'

I'm sure the beer sales of these clubs can never be the same since the steam firemen ceased to exist.

9 · The Locomotives

On reflection I can never recall any GWR enginemen describe their locos with such glowing intellectual detail as is the case today.

Expressions such as 'Prairie' I never heard. Tank engines were either side tank, saddle-tank or pannier tank. A 'Mogul'? Well, most of us thought that was a Mongolian!

Barry was essentially a tank engine depot with the exception of holiday times, when visiting tender engines from a dozen different sheds took predominance, while their passengers enjoyed a day at the trippers' paradise of Barry Island.

The standard of locomotive husbandry at Barry was extremely high, having, of course, the Barry Works near at hand, where members of the active Mutual Improvement Class had *carte blanche* to visit. But, going back to locos, they were basically a tanky or tender, with more emphasis on their company of origin. The locos of the Taff Vale Railway were 'Taffies'; those of the Rhymney Railway were 'Rummies'; the Barry Railway locos were affectionately 'Old Barrys'; and GWR locos were 'Western Engines.' The later British Railways newcomers were scorned as LMS mutton done up as lamb, or LMS 'Bagpipes' which lacked the pipe-free lines of the Western engines.

Nearly 100 locos were housed at Barry, all tank engines, ranging from the tiny ex-Barry 0–6–0 up to the powerful 7200 2–8–0 tank class. With a large dock area to serve, a good third of the work was taken up in pilot duties, shared by the old workhorses of the Rhymney, Taff Vale and Barry railways, and backed up by the toughest little engine of all time, the Great Western 5700 class pannier tank. But if the workload of the pilots was any criterion, the lot of the passenger and mainline locos could be described only as herculean, for there are no easy straight runs in South Wales.

Except for very rare changes, ex-Taffy tankies did most of the passenger work. These were 0–6–2s with 5ft 3in. diameter coupled wheels. They maintained a faultless service between Barry, Bridgend, Cardiff and the valleys providing the coal was right. Anything else than good Welsh large, and they disapproved vehemently on their steam pressure gauge of a normal 200 lb. At a later date about half a dozen 5100 class 2–6–2 tank locos appeared on the scene to replace some of the ailing 'Taffies' who made the terminal journey to Swindon Works for scrapping. On one occasion I fired on such a journey and despite the perk of an easy job, I had the guilty feeling of taking a faithful old barren cow to the slaughter-house.

I suppose the general 'Jack of all trades' of Barry Shed, and indeed all South Wales, was the 5600 class, designed specially for the South Wales coal trade, which they handled admirably. They also stood in for branch passenger work, and dock pilot duties. The '56' was worked to her limit and more. Many is the time they have started away from Cadoxton Yard for the valleys by virtue of a real blow-off from the safety valves, not just a strong white feather, but a screaming banshee that pierced the eardrums. It used to be a combination of nail biting and prayer, not to mention slave labour, to help those brave locos come to grips with the torturous gradients of South Wales.

Unlike the fickle 'Taffies', who had changed to GWR boilers, the '56s' would take most types of coal in their stride. When I looked into the firebox of a fresh 'Taffy' to see a clean box, a new brick arch, then turn around to a bunker of mediocre coal, I would wish for a '56' with a bunker of small. It was a constant unsolved mystery to me that two quite similar boilers should behave in such contrasting manners.

But that is what the job was all about – no two engines were alike, and even the veteran enginemen could not supply the answer. Some would steam on a cigarette lighter, others were 'firemen killers' from the day they were built.

To this day, during a life of many ups and downs, I have yet to suffer a more nerve-racking experience than firing a bad steamer. Just imagine a 'Taffy' fresh from the shed, waiting at Barry station to pull away with six coaches for Bridgend. The boiler is full, too full for good engine working. The clean fire slopes from half-way across the ring to about 9 in. at the front of the box, there is a little smoke, and the steam pressure gauge rapidly reaches the red mark of 200 lb.

The porters are shouting and checking the doors, the guard blows his whistle, and the 'Taffy' is screaming at the safety-valve with impatience as the steam from her full boiler showers a fine rain along the platform.

'Right away, mate.' The cylinder cocks have to be opened to save the front ends of the cylinders being blown out. The driver moans about 'bloody hydraulics' and the quivering needle of the steam gauge begins its descent. The driver looks anxious as he shuts off to bring the lever back another notch during the hard twisting climb of the Vale of Glamorgan 'switchback'.

She'll only get worse with the slightest splash of coal, demanding an incandescent fire. The fireman draws the pricker from the tank top as she enters Porthkerry Tunnel, and pulls his cap over his eyes against the white hot heat as he rips away at the fire. She's holding on to 180 lb on the clock as she banks through the hole.

Suddenly there is daylight and she's crossing the viaduct, but the once-full boiler is down to a third of a glass, so the injector must go on. It won't be long, though, until she can free wheel down into Rhoose station.

The fire is a foot below the ring, but it's safe to put some on, only on the back and let it dribble to the front!

Same thing again at Rhoose, blowing her heart out, full boiler, red fire, but she wants coal like a blast furnace. 'Right away, mate,' downhill to Aberthaw station. She's screaming from her safety-valves fit to burst. Now it's an uphill slog for 6 miles, so soon her safety-valves allow only a little white wisp to escape, then silence as her fireman fights to maintain the all-important 180 on the clock.

That's how a 'Taffy' worked the Vale to Bridgend, demanding every ounce of effort from driver and fireman.

The same journey with a '56' would pass without effort, yet every time a 'Taffy' was finally withdrawn from passenger work to work on dock pilot duties prior to her last trip to Swindon, there would be an outcry from her engineman.

I sometimes wonder if all footplatemen were masochists at heart, or conversely, born fighters, for they appeared to either love the challenge of a 'rogue loco' or be ruled by her. I would often listen to a heated debate in the mess-room between ex-Barry men and Western men regarding the merits of their various locos. That was during the 1940s, and my young mind failed to comprehend the comparative short distance back to 1922 when the small companies were amalgamated into the Big Four.

The Barry men spoke affectionately of the 'L' class or 'Hell' class, I never knew which, but it could have well been the latter when they described some journeys.

The Western men from other depots usually quoted 'Castles' and 'Kings' with a superior air, to which the Barry men were quick to reply 'We had bloody tenders, too.' Then the Western men would keep it going with 'Well, where are they now, couldn't handle 'em could you?' I suppose a good third of my firing career was spent within the 'Spare Link' where anything could happen.

Occasionally I was dispatched to Cardiff (Canton) Shed with tongue in cheek, for Barry could not be compared with such an important mainline depot. The locos ranged from 'Castles' down to little 4800 class 0–4–2 auto train fitted tankies. During my early days these pretty little engines mystified me as they passed by on the Penarth branch line pushing their train. The fireman would be working out of sight, and with no driver, the train appeared to be going on its own.

Canton, it seemed, had the lot except 'Kings', which in those days were never allowed beyond the English side of the Severn Tunnel, due to various restrictions, but I understand that they did eventually arrive on the scene during the 1950s.

Canton drivers were devoted 'Castle' men, however, speaking of their cherished locos in more than holy terms, and on the rare occasions that I fired them, I felt I worked with an owner driver.

At a later date when British Railways introduced their LMS style equivalent to the 'Castle' – the 'Britannia' class – I could well commiserate with their feelings towards 'Bagpipes'. The 'Castle', in her shining green livery, had the smooth sleek lines of a thoroughbred racehorse, whilst her new BR counterpart was decidedly opposed to any resemblance of GWR design.

There was a lot of bad feeling on the part of Western men at this period of nationalisation. They felt that a fine company such as the GWR had been completely ignored regarding locomotive design, and many feared that the automatic train control, and indeed the whole of the Western signalling system might change in favour of LMS procedure.

Canton also had its share of ex-Taff Vale and Rhymney locos, usually employed on pilot duties, many of which were built well before the turn of the century. The war had provided a reprieve for them, and with little in the way of extra maintenance, they carried out their daily duties with no more than a grunt and groan.

I don't know if diesels have characters or hearts or spirits, but steamers most certainly do.

Take any engine shed about two in the morning – perhaps sixty locos under one roof – all in various stages of steam – awaiting their turn to go out to their different tasks: proud passenger engines, which appeared to be aloof to the humble pannier tanky at their side; powerful mainline goods engines, such as the '2800' class, which seemed to flex their steel muscles to all and sundry.

Once the fire was lit, locomotives became alive. Their hearts began to beat – a kind of blood flowed through them, and they displayed their characteristics.

I knew of a pannier tank, a '57' class, which decided to do her own thing one very warm mid-morning in August. She had travelled through the reception services of the shed after working three shifts as a dock pilot. She became weaker and weaker in steam, as her fire dropped, and she moved on to the coal stage. The final operation was to drive her into the shed, a duty performed by the Bottom End shunters, two firemen and a driver.

Now the Rule Book said that these gentlemen, together with their own engine, should place each coaled loco into the shed. But, of course, rules are made to be broken, and if, in their expert opinion, there was enough steam, they would drive the loco to the shed and not push her.

Next, however, we broke another rule. Two men should be on the footplate, while the third held the points. Quite often only one man would take the engine to the shed. This was a game of chance that rarely went wrong, but it was essentially for a young man, as he had to be fleet of foot.

The method was to take the engine, say two lengths beyond the set of held-up points, set the regulator so that she only just crawled forward, then run and hold the points, jumping aboard as she went past. It had never been known to fail.

Eleven o'clock in the morning was a quiet period on most sheds. The earlier hustle and bustle had calmed, and a type of anticlimax reigned before the start of the midday rush. I was sitting outside the enginemen's mess-room on spare duty. This was the type of job where you could be worked to a standstill, or do precisely nothing for eight hours, just killing time playing cards or dominoes, and consuming gallons of tea.

The pannier tank came out of the shed on No. 3 road at a fairly brisk pace for a shed movement, perhaps 8–10 mph, followed by excited echoes from within – 'Stop her someone! For Gawd sake, stop her! She's a runaway!' said a breathless voice. Soon the red-faced figure of Steve Rodgers appeared, distressed and hatless as he tried vainly to catch the elusive engine.

The scene suddenly changed to a comical fairy-tale situation, as fire-droppers, steam-raisers, fitters, boiler-washers and every ancillary shed staff possible joined in the chase. The strange thing was that no other locos, or traffic, were in her way, as she puffed merrily away from the shed, followed by her large band of pursuers showing various states of fitness. But without a fire the gallant pannier was fast losing her breath against the shed incline, enabling a far from swashbuckling Steve Rodgers to reclaim his prize.

Later in the mess-room, to a sarcastic audience, Steve related how it happened. Apparently, when he set the engine to move forward before running to hold the points, he had quickly put the gear lever in the 45 degree position, which would ensure that the engine would be quite sluggish in moving away. However, as he held the points, he could tell by the sudden heavier blast and faster movement that the lever had slipped out to its full extent.

Luckily for Steve the matter was not referred beyond the shed. He received only a local caution, and, despite his declaration that he would work to rule from then on, he was back berthing engines single-handed a few weeks later.

Another pannier tank, however, seemingly left the shed one evening to return just before dawn without a crew, but I happen to know the secret of that legend.

'Let's go dancing in Cardiff,' said Ginger in the Railway Club one night. 'I'm fed up with the local hops.'

'So am I,' said Buster. 'But if you take a girl home you miss the last train, and that means a twelve-mile walk home.'

Ginger, after a long draught from his pint, and a serious session of deep meditation, spoke quietly into Buster's ear. 'Borrow a loco, mate. That's the thing to do.' Buster laughed at the idea, but Ginger, a renowned and fearless practical joker, was serious. 'You know what it's like on the shed at this time on a Saturday night. As quiet as an undertaker's.'

Buster's staunch caution remained, but later, with the courage of a further pint inside him, he reluctantly followed Ginger to the shed.

It would have been a strange sight to even the most unobservant – two smartly dressed young men quietly boarding a pannier tank near to midnight.

Ginger, a mainline fireman, smiled at his carefully selected plan, as he quietly eased the loco off the shed. She had to have a steam brake – the ejector of a vacuum brake engine would be too noisy and arouse the few night staff on duty. His wide knowledge of loco shed management was also put to good use, for he had to select an engine that would not be used for a few hours. The one he drove off Barry Shed was booked to the permanent way department to head a Sunday track relaying train.

The signalman at Barry Junction allowed Ginger and Buster onto the mainline without a moment's hesitation when they phoned him from the shed box to say 'Light engine to Canton Shed.'

With all signals set and the message passed from box to box, the little pannier arrived on Canton Shed within half an hour, and it was then that her illegal crew had to toss the coin. Having little or no knowledge of the shed, they decided on a short siding to berth their taxi, before tripping away merrily to the nearest dancehall.

During the early hours of Sunday morning, the two late-night revellers returned, after an enjoyable evening, to the exact spot on Barry Shed, and quietly sped home to their beds, and no one, except an elderly steam-raiser, was any the wiser. In fact, no one believed the poor old man when he spoke of the strange goings on of a ghost engine that night.

His list of departure times read 7.30 a.m. for the pannier. 'She was right there on Number Six road when I signed on at ten,' he said. 'I puts a bit of coal on just after that, but when I goes to check her about two o'clock she's gone! Then, blow me if I don't see her coming back onto the shed straight down Number Six as if she'd never left the spot!' The old man took a drink from his cold tea bottle before continuing his story to a sceptical gang of coalstage labourers. 'When I goes over to see where she'd been, there was no one on her, and when I goes to the enginemen's mess-room, I sees only a fireman and driver fast asleep.'

10 · Down the Vale

The Vale of Glamorgan branch line was typical of the rural transport system of the Great Western before road transport sounded its death knell. That picturesque line between Barry and Bridgend is now rarely used, only as a mainline diversion during line repairs and for coal delivery to the power stations at Aberthaw.

During its heyday, apart from the hourly passenger service to Bridgend, the line was used by countless coal trains from the Western valleys bound for the Barry Docks. The regular little busybody was duty B15, always grossly underpowered by one of the ex-Barry 0–6–2 tankies, for no other reason than habit. All the South Wales sheds were notorious for protocol, and, wherever possible, the locos, which worked their various jobs before the 1922 grouping, continued to do so with meticulous routine until their final call to Swindon scrapyard. Unfortunately, the boiler pressure of these elderly ladies was constantly subject to reduction, varying from 140 lb on the real old-timers, up to 175 on the more healthy.

Canton Shed, for instance, kept an old ex-Taff Vale Railway loco for ever and a day at Ely Harbour. Even in those days, twenty years after the death of the TVR, the Company lived on. Nothing had changed from the coal hoists, the permanent way, the signal boxes, the Company signs, and, of course, the faithful old loco. The little old world of Ely Harbour remained distinctly TVR, while surrounded by a

Great Western system, and as far as B15 was concerned the Vale of Glamorgan line was still the Barry Railway. From Barry to Bridgend the old loco wheezed her way through the tunnel, over the beautiful viaduct to Rhoose asbestos works, her first call. For an hour or so we would shunt hard, delivering the inwards traffic such as coal fuel, engineers' spares, and raw materials, finally collecting the train of finished products: corrugated asbestos sheets. But these were only taken to a convenient siding for collection on the return journey. A few miles, and as many gradients, along the line was Aberthaw cement works where the procedure was repeated, and wagons of cement put aside for the homeward trip.

Of course, at each station there was a siding for the local coalman to be shunted, and most farmers used the system for delivery of new implements and livestock transport. There was even a line running into the huge RAF camp at St Athan where we delivered and collected everything from food supplies to aircraft, one of thousands of lines devoted to private users. Many ran right into the heart of factories, providing a service that was second to none. To this day I fail to compare the total volume of production on such a train as B15 with road transport, when I consider that three men and a veteran locomotive could provide virtually all the transport requirements for the Vale of Glamorgan. Goodness knows how many lorries are required to carry out the same work these days.

Llanwit Major was roughly at the half-way point of the Vale journey, a quaint old town many centuries old and with a proud boast of nearly as many pubs as houses. But more importantly it was where we made the tea.

To take our official twenty-minute break was unheard of, although outwardly we were far from devoted 'Company men', but B15 allowed us no time to stop, so we had to eat and drink on the hoof. On then to Wick, Southerdown

Road and finally the savage curves of Ewenny. Most of these had restrictions of 25 mph which were completely ignored by the passenger men in order to keep time. One driver who terrified me on that part of the line was Cleo Gill, a small wiry man who had a paranoiac aversion to noise and coal-dust. He would sign on well before time to 'insulate' the engine, using most of our cotton-waste allowance to plug the various gaps between the cab and the boiler which usually created quite a rattle. But his dust-prevention routine was most meticulous. As soon as he arrived on the footplate, the injector would go on and the pep pipe would seek out everywhere within range, from the coal bunker and cab top to footplate, tanks and boiler. Then he would produce his elastic bands, half an inch wide, stitched in a circle, to close his overalls tight around his wrists and ankles. Most of the firemen left the footplate hygiene to him for they could never spare the time, but although there was virtually no dust about, there was plenty of water and wet feet.

Bunker-first journeys usually resulted in the pep pipe running non-stop, and that brings me back to the curves of Ewenny, which twist through corridors of hewn rock-face. One way or the other we had to travel bunker-first, and with Cleo's obsession of dust firmly embedded in his mind, he forgot the curves which we would strike with a frightening lurch at up to 60 mph. As I said earlier, most drivers ignored the speed restrictions, but none with such contempt as Cleo Gill, who was more than surprised one day to see the tailend of the local hunt as he rounded a bend in the seemingly endless cutting. Cleo was a time-keeper *par excellence* and no hunt was going to smear his unblemished record. With cylinder cocks wide open, the ex-Taffy closed on the terrified hunt like some hideous monster.

As Cleo said later in the mess-room, 'I reckon the horses wouldn't have come to shame in the Grand National, and the hounds would have won the Greyhound Derby.'

At its western end the Vale line ran into Bridgend station on the GWR mainline, where B15 deposited the last of her outward train, and collected her up goods, as the gleaming 'Castles' of the Paddington–South Wales expresses passed by. With the exception of the regular coal train service to Tondu, and a few special passenger trains to Porthcawl, little else from Barry Shed worked beyond Bridgend, and so the old ex-Barry Railway tanky ran round to repeat her outward journey and head for the Barry Sidings, at the western end of the town, a quite large marshalling yard which handled the busy flow of traffic bound for the Vale.

It was the duty of the yard shunter to be 'banker' engine to assist goods trains through the inclined Porthkerry Tunnel at the start of the Vale journey.

My driver, during a week's stint on the Barry Sidings pilot, was Albert Bugler, with the nickname of 'Dapper'. He was best described as a Marxist Capitalist, a strong trade-unionist, with an ardent fervour to help all his brothers – at a price! He was also a keen gardener of the railway embankment allotments at Barry Sidings. One of his horticultural successes was the production of tobacco – a deep brown, long-burning concoction, known to his suffering customers as 'Dapper's Dark Honeydew'.

Our head shunter was Jack French, a powerfully built West Countryman, who was obviously strong enough to consume a regular supply of Dapper's duty-free tobacco, despite the warnings of his mate; Bill James. Bill was a bookworm, a ten-ticket member of the local library, who read anything that came his way.

'You could have fits smoking that stuff,' he said, while Dapper sold Jack his weekly supply as we hauled a mixed train for shunting.

Jack shrugged his shoulders through a haze of smoke. 'Trouble with you, mate, is those blasted books! You don't want to believe all you read. Turn your head it will.'

Bill was insistent. 'I'm quoting an excerpt from the *Lancet*, mate, an' if they don't know, who the hell do? Saltpetre's the trouble. Too much of it and you go bonkers. How do Dapper know how much to use? You tell me that then!'

'Rubbish,' said the burly Jack, as he lowered himself from the footplate on to the wooden coverings of the signal rods below.

Bill ranted on to the point of being a nuisance, as we steamed on awaiting Jack's signal to stop. After covering a distance of about thirty wagons from Jack, I noticed that he was dancing and throwing his arms in the air like a Red Indian, brandishing his shunting pole like a spear.

Bill James momentarily went silent, then burst into an excited tirade. 'He's got one! He's got one! What did I tell you? Too much saltpetre. Damn, I can see him frothing from here!'

Dapper, who had constantly ignored Bill with utter contempt, rushed over to my side, as the loco steamed on still awaiting a hand signal to stop. Two figures appeared to be restraining Jack, as a third waved his arms to give us the long-awaited stop signal, then call us back into a siding.

Bill's voice quivered with excitement. 'They won't move him from there without a strait-jacket. I'll bet he's got the strength of three men.'

As we drew nearer to the dancing figure of Jack and his helpers, their voices indicated something totally different from a fit. 'Cut his trousers and get it out,' said one. 'Let it go and shake your trousers,' said another. 'It might run down your leg.' The third man, Chepstow Jack, a platelayer, was more daring in his tactics: 'Stand dead still, Jack, an' I'll belt him with this brake stick.'

As we arrived on the scene, Jack had cooled down and was holding on tightly, with both hands, to an object beneath his trousers. Slowly he unbuttoned his fly-front with one hand, as he continued to hold the object with the other. The scene was dramatic, as the white-faced shunter slowly released button after button before his stark-faced audience. Suddenly, the head of a large rat appeared, and, as Jack released his grip, the animal scampered down his leg to hide under the nearest duckboard.

Later, during our food-break, as we all discussed the incident, Jack explained how the rat ran up his leg when he first jumped from our engine onto the signal rod duckboards. 'Must have disturbed the blighter,' said Jack, as he sipped a welcome cup of tea. 'The first ruddy hole he bolted for was my trouser. Ruddy good job I wears long underpants. Could have been serious, it could!'

Bill James sat morosely in the corner of the cabin pretending to read a book on the life-cycle of spiders, as Dapper Bugler shouted across to him, 'You want to forget about saltpetre, Bill. Read a book about rats an' we'll know what to do next time!'

Our break was short-lived, when the yard foreman came in. 'Get in front of the Llandore, lads, and assist her through to Rhoose. She's just one wagon over her load.'

She was one of the powerful 7200 class 2–8–2 tank engines with an extended bunker. Her Rogerstone-based crew gave us a superior grin, as our ex-Barry Railway tanky coupled up to head the train. As we entered the Porthkerry Tunnel, it was obvious that the '72' was taking advantage of us, for, while we were literally blowing the roof of the tunnel off, she was working on half regulator. Dapper blew hard and long on both whistles for more effort, but his call was ignored by the driver of the '72', who was obviously out for an easy trip at our expense.

We could, of course, have eased off our own regulator, but that would have meant sticking in the tunnel, and no footplateman relished such a choking experience.

Dapper was furious. 'Pull the smoke-plate out and pee on it, mate,' he shouted into my ear.

I had heard of such treatment for lazy engines many times but passed it off as old drivers' yarns.

'Come on, mate,' said Dapper again, 'we'll be out of the tunnel before long an' we'll never get 'em moving.'

I gingerly withdrew the red hot plate on my shovel to reveal a white-hot fire to my tender anatomy and peed. The response was electric: the '72' surged forward, barking her exhaust out with a full regulator to escape the obnoxious gas of the evaporated urine.

Dapper jumped for joy, as I hastily covered up my scorched skin.

'I'm not doing that again, Dapper,' I said, as we headed back 'Light Engine' for Barry Sidings. 'I'd rather choke in the tunnel than suffer it roasting.'

'No need to, mate,' said Dapper, 'I'll be ready for them tomorrow, if they try it again.'

He told me later how he had spent his early driving days on the Severn Tunnel 'Bankers' and peeing on the smoke-plate was considered a major part of the fireman's duty. Bill Edwards, the shunter, heard this and quipped back 'I don't know why you had to go to all the trouble of peeing. Why didn't you light up with a pipeful of your Dark Honeydew? It smells just as bad!'

The following day, Dapper arrived carrying a large canvas bag. 'Shove them outside on the coal bunker, mate,' he said, handing me the vile-smelling bag. He rubbed his hands excitedly, as I held my nose from the obnoxious smell. 'That's the real Mackay in that bag, mate! Beats peeing. That's called the "Banker's Blessing", mate. A whiff of that, only a whiff, and a lazy train engine will steam to hell and back!'

Later that day, returning to our loco after break, I paid little attention to the hundreds of screaming sea-gulls which circled the engine, for we were a coastal town. Dapper could scarcely conceal his impatience for the arrival of the Llandore goods, and, when eventually the '72' laboured into the yard, he chuckled across to me: 'Fresh crew, mate. If they don't pull their weight – look out!'

As expected, the Llandore requested a pilot and, as we coupled up, I began to half-wish they were a lazy crew in order to see the effect of Dapper's mysterious evil-smelling bag. They carried out the usual bluff at the start by giving the '72' a full regulator, then slowly, very slowly, easing off, so that we were forced into giving our all to keep the train moving.

At the centre of the tunnel, Dapper tapped my shoulder. 'Get the bag, mate!'

I opened the bunker doors to heave it onto the footplate. It was empty! Picked clean, I realised, by the voracious sea-gulls. Dapper's face fell in the light of the white-hot firebox when I told him of the birds. 'They would do, mate,' he said. 'It was a bag full of fish-heads. Rotten fish-heads. Sling 'em just inside the firebox you do. Worse than any poison gas.'

I lifted the bunker slide to start firing, when he stopped me. 'We can't let him get away with it, mate, pee on the smoke-plate!'

I was aware of my bodily limitations. 'I can't do one,' I said.

Dapper was persistent. 'You've got to, mate. We'll be out of the tunnel 'fore long.'

'I can't,' I repeated, 'there's none in me.'

He stepped into the well of the footplate to lower the smoke-plate, then unfastened his trousers. As he removed the smoke-plate to receive its douche of unholy water, a boiler tube burst, scalding his penis.

I found it hard to suppress my laughter, as he danced about the footplate, and for the remainder of the week we suffered in silence the continued laziness of the Llandore goods.

11 · Meals on Wheels

Steam footplatemen generally enjoyed very good health and longevity, despite the lack of a good meal while on duty. On dock pilots or shunters, the permitted break was twenty minutes, while on mainline goods or passenger work, food and drink would be snatched whenever convenient. There was a marked contrast in the appetites of a driver and a fireman. Most drivers were content with two or three neatly cut sandwiches and a piece of cake from a small 'tommy tin', but the fireman's food consumption was enormous.

There were exceptions, of course, and 'Tom the Rag' was one such engine driver. His real name was Thomas Davies. I don't know how he acquired such a distasteful nickname, and no one else did for that matter. He was a tall, gaunt figure with a long sinuous neck, and drooping moustache. He carried a legend as the only man who could keep steam on No. 90, a long-since scrapped loco of the old Barry Railway Company.

It was a known fact that if you required anything, within reason, of course, from, say, an odd size nut and bolt to a broody hen, Tom would have it. One fireman related how he called at Tom's house for some oddity. The rooms were packed solid with the spoils of a lifetime of hoarding, and as he sat around a cramped table for a cup of tea he could hear the sound of a hen and chicks beneath it.

I first fired for Tom on High Level, where he drove one of the many coal hoist pilots which humped their trains of black diamonds into the hungry holds of waiting cargo ships for transportation to all parts of the world during the reign of King Coal.

Tom carried his food in a fish frail – a large woven rush basket. He called it his nosebag. Throughout the shift he ate from it as he drove the ancient tanky to and from the coal tips.

As soon as the shunters left us for the glorious twenty minutes of food and cards in their dockside cabin, I took water, whilst Tom meticulously cleaned the pan of my shovel with a clean piece of cotton-waste.

I was to enjoy a large Spanish onion with a slice of Cheddar and half a loaf of bread. The onion had been baking behind the water injector clackbox since the start of the shift and only its skin kept it intact like a bag of jelly.

I watched Tom carry out his role as 'Chef de Cuisine'. I could hardly believe my eyes when he poured a quantity of 'Blackjack', the thick, dark green cylinder oil, onto the shovel. With the stage flourish of a magician, he produced two large 'bloaters', or smoked herrings, from his frail to join the unholy cooking fat. The smell was obnoxious, to say the least, as they began to cook in the mouth of the firebox, but he ate them with the same relish as smoked salmon or caviar.

Tom obviously misread the dumbfounded look on my face as one of envious hunger for his meal, saying, as he held the shovel in my direction: 'I see you got a bit of bread left, mate. Dip it in the fat. Go nice with your cheese and onion.'

Of course the footplate frying pan, the fireman's shovel, has always been renowned for its famous fry-ups, but I recall a situation when I came near to the murder of my driver. We had worked a 4200 class 2–8–0 tanky on a night coal train from Barry to Bridgend – empties down line and coal back. At the Tondu Sidings, the guard uncoupled our train for us to stand by outside the yard foreman's office whilst taking a short food break.

Bill Bouchier, my driver, a wonderful personality who remains dear in my memory, joined the guard and the foreman while I prepared my once-monthly fry-up. It was post-war, but food was still rationed and in short supply. Unlike Tom the Rag, I washed the shovel with the boiler hose. No frying pan from the best household in the land could be cleaner! With a stomach rumbling with hunger and excited anticipation of the feast to come, I delved into my big black 'tommy tin' – two eggs, half a pound of streaky bacon, a ring of black pudding and a large basin of bubble and squeak. No engine oil for me either. A lump of beef dripping was supplied to lubricate the shovel. During the ecstasy of preparing my meal, I was unaware of the labouring approach of a '72' class tanky, as it hauled our train of coal into the same siding we were standing on.

As I gave a final turn to the bacon and returned the shovel to the firebox, Bill Bouchier clambered up to the footplate, taking off the handbrake and blowing up the vacuum brake ejector almost at the same time. 'Our train's coming in, mate. We got to get out of here,' he said, as he opened the regulator.

I was too slow in withdrawing the shovel as the first blast on the exhaust sucked my glorious spread to an ignominious end. With great difficulty I suppressed the desire to beat Bill to death with the empty shovel. Our footplate atmosphere was frigid for a few days after that unfortunate episode, and perhaps it may have been a pleasant release for my poor stomach in not having to contend with that shovelful of greasy food.

A few weeks later I had my first taste of express passenger work. I was sent to Cardiff (Canton) Shed as a spare fireman and, to a mixture of horror and tense excitement, I replaced the regular fireman of the evening express to Paddington.

The driver was Bill Bousie and the engine was 'Castle' class 5052 *Earl of Radnor* – a fine engine and an equally fine driver. A 'prep' crew had got her ready to join her train and taken her to No. 2 platform of Cardiff General station. At that stage, with about twenty minutes before pulling away, there was absolutely nothing to do. The firebox was full, as the steam pressure needle quickly crept up to blow off. The tender was trimmed and damped. The wooden footplate was as clean as a butcher's block. Like all the beautiful thoroughbreds of her lineage, she was raring to go.

Bill Bousie handed me his enamel tea-can and the little oval tea and sugar tin. 'Nip back to the restaurant, mate, and ask the chef for some hot water for a brew.'

I felt quite important as I passed our passengers on my way to the centre of the train, but steam locomotion was the accepted method of transport then and I don't suppose they took a second glance at me. I tapped the louvred door of the chef's coach.

'What yer want?' said the faceless voice behind the shutters.

'Can I have a can of hot water for a brew, please?'

A volley of abuse soon deflated my short-lived ego. 'Ain't yer got enough blooming hot water on up front without coming back yer to nick mine?'

With an empty can, and much less pride, I returned to the engine. What was all this conceited talk of being the cream of the railway!

Guards call it *their* train. Signalmen said *their* stretch of line and *their* box. Even the catering staff placed a privilege on *their* precious hot water!

Bill Bousie appeared unperturbed. 'That's O.K., mate. Make it from the boiler. I'll never know why we're so fussy. I don't suppose you'd find a more hygienic drop of water. It's super-heated.'

Within a few minutes, to the shrill of the guard's whistle

and my own pounding heart on my first express trip, we pulled away with fourteen packed coaches. Twenty minutes later we ran into Newport station, fiercely blowing off, despite closed dampers and both injectors working. The stationmaster in his fine regalia, who appeared only for expresses, momentarily shook his head, holding his hands to his ears, for I was breaking the holy rule of blowing off in stations.

Soon we were heading for Severn Tunnel, with the famous 'little and often' routine of good Welsh steam coal to the white-hot firebox. Boasting a full boiler, and a little white feather on the safety-valve, it appeared all too easy, but, to date, it had been flat or down gradient with a clean fire, and Paddington was quite a way to go.

The long haul began roughly at the centre of the tunnel, and about twenty miles and a half-ton of coal later I was a bath of sweat looking forward eagerly to the long flat stretch of Patchway water troughs.

My inexperience in handling the scoop caused a flooded footplate, but my good-humoured driver passed it off with a broad smile, as we picked up speed for the easy run through Badminton. Soon we were in the top seventies, with just a quarter of a regulator and the gear indicator well back, when I noticed Bill Bousie wind it back still further. The loco began to take on a backward and forward lurching movement, with such a short stroke on her valve gear. I found it difficult to keep steady on my feet, and a few shovelfuls of coal hit the side of the ring. I blamed myself for lack of experience, and I was quite relieved when, twenty minutes later, Bill let the lever out a little.

We made Paddington dead on time, and I don't think the station was large enough for me as we steamed proudly toward the stop-block.

Bill handed me the tea-can once more as soon as we came to a standstill. 'Try him again, mate,' he said. 'He may be in a better temper now.'

I walked back gingerly, expecting another verbal onslaught, but the door opened to my knock to reveal a rather flushed-looking chef. He took my can, refusing the tea and sugar tin.

'Here! Have one on the house. Sorry about being so edgy at Cardiff.'

I thanked him and turned toward the engine.

'By the way,' he said, 'tell your mate we all got the message.'

I was completely puzzled over his change of attitude. 'What message?'

He called back to me, with a nod of his head. 'Your mate. He knows bloody well when we're serving dinner, don't he?'

He was aware of my innocence.

'Badminton, mate! Didn't you notice the motion going through Badminton? We were just serving soup. Couldn't keep the bloody stuff in the plates, we couldn't.'

Bill Bousie gave a broad smile as I returned with the free can of tea and the chef's compliments.

I didn't work the Red Dragon anymore that week, but I'll bet the regular fireman had no trouble in having his tea made at Cardiff.

Tommy Twigg was a dock pilot driver who had the cheek of 'Old Nick', and the ships visiting Barry were victims of his 'hot water' ruse. It was wartime and vessels of neutral and friendly nationality called at the port with well-stocked food galleys. Tom had a smooth tongue, and an expert knowledge of flags. Greek and Turkish ships were avoided for their lack of generosity whilst Scandinavian were always approached for a can of hot water to make the brew.

Very few ships' cooks could resist the humble advances of Tom as he made his false request. 'Put your brew away,' they'd say. 'Here, have a can of ship's coffee, and some cooked ham.' And Tom would later come down the gangway, his face beaming and his arms full of spoils whilst the devious brew kept its virginity for yet another day.

But the ever-generous American ships were top of the poll for Tom's attention. Their huge stainless-steel galleys drew him like a magnet as their cooks prepared cuisines well above the average merchant ship standard.

No one could refuse Tom, his full rosy red face portrayed a constant feature of goodwill and well being, as I followed him aboard a liberty ship during a meal break. As he homed on the galley like a pigeon the two cooks on duty looked up from their work. 'I wonder if I could trouble you gentlemen for a drop of hot water to make my tea. Haven't had time to spit since we signed on, and the war comes first I s'pose.'

I mused on how the war stopped us from making a brew as Tom continued.

'My goodness, there's a sight I haven't seen in many a long day, white bread. You bake it yourselves, lads, do you?'

Before Tom could save his precious brew one of the cooks took the can and threw the contents into the sink. 'Hell, man, I don't reckon us Yanks can do enough for you Limeys. You've had enough of that gawdamn gnats piss, just you try some of Uncle Sam's good ol' US hundred per cent coffee.' Tom looked on in a mixture of despair over the disposal of his record-breaking brew and gratitude for the impending wealth of goodies to follow as the emotional cook continued: 'Why hell, man, just you help yourself to a few loaves, and here, take some tinned ham, and salmon, and chocolate and –'

Tom was going for the complete kill. He produced about a third of a flattened cigarette end from his waistcoat pocket. 'You're very generous, I must say. Could I trouble you for a light, matches are another scarce commodity in the old country.'

I began to feel a flush of embarrassment, for Tom was

working his way down to his boot laces, and half an hour later we struggled ashore with our spoils, which included cigarettes and a lighter.

His personality was endearing, and quite a number of crewmen would call at his local pub during return trips to the port. One ship stayed at Barry for several weeks during repairs to her main engines, and Tom, apart from becoming completely impervious to British ration books, also cemented a strong friendship with her chief engineer.

Tom was working a New Dock pilot on the day that she sailed. With his usual affectionate manner he embraced my shoulder as we pushed a train of mixed goods along the dockside. 'Be a pal, mate, and take the regulator for ten minutes while I nip aboard to say cherrio to my friend.' It was an obvious rule-breaking offence, but experienced dock pilot enginemen could be quite skilled in working their engines single-handed. Once the fire was well banked up they would move from one side of the footplate to the other, as the train slowly snaked its way along the fierce curves of the dockside.

I didn't miss Tom for his first hour of absence, and when the tugs arrived to take the ship to sea I expected his return to the footplate, but a very busy session of work took my mind completely off the subject.

'That's it, then,' said the head shunter a couple of hours later. 'Let's have some grub.'

His eyes scanned the footplate as he looked up from the four foot. 'Where's his lordship then, as if I didn't know?' I looked across to where the superstructure of the American freighter had always been in view. Only the jibs of the dockside cranes broke the skyline. The shunter read my thoughts. 'Quick, borrow the foreman's bike and nip around to the locks, I 'spect they're having a farewell booze-up, and forgot all about the sailing.'

With no time to ask for permission I pedalled the cycle

furiously along the dock road, only to see the grey painted stern of my quarry leaving the dock entrance. I didn't quite know what to do as I watched the ship become more minute in the low afternoon sun of the Bristol Channel. How does a fireman tell a shed foreman such as Mr Chidgy that his driver is on his way to the United States, I thought. Both the shunters and myself treated the affair light-heartedly as we discussed it on the footplate of the little '57' after meal break.

'She's well down channel by now,' said the grade three shunter. 'They won't turn round with a ship that size even if they do discover him. Besides, I s'pose they got to join a convoy on time.'

The head shunter laughed. 'Tom would make a good Yank anyway, he's got enough cheek for one, but seriously, what are we going to do about it? Someone's got to be told sooner or later.'

Various suggestions were produced by the three of us as the shift wore on, and the moment of truth grew menacingly near. Our last orders were to take four brake vans to Barry Goods Yard, and then on to the shed. 'Well really it's nothing to do with us,' said the head shunter. 'We're traffic, and you're motive power, so the ball's in your court, so to speak.'

Suddenly there was the unmistakable high-pitched ring of Tom's voice: 'Anymore for the Skylark?' He looked up from the four foot.

'How did you get back?' I said.

His face beamed. 'On the pilot, hell of a shock it was. Having a little drink with one of the engineers when I feels the deck moving beneath me. "Hell," I said to my pal. "What you got in my glass then?" Just got the pilot in time when we realised we were at sea. Blimey, I got a head on me. Let's have a brew-up before we go on the shed.'

12 · Chicago Stockyard at Barry

Because of its strategic position near the Western Approaches, Barry became an obvious choice for the American Forces during the Second World War. The port flourished, as it catered for the constant flow of liberty ships, which carried everything from toilet tissue to tanks.

Upwards of twenty dock pilot shunters worked night and day to handle the fuel of their gigantic war machine. Main-line engines and vans would wait for as long as four hours, while the busy pilots arranged their trains for various US camps throughout the country.

The Americans of Barry also had their own marshalling yard, a 500-acre wasteland, converted in a matter of weeks to a replica of a Chicago stockyard. Encircled with a 9-foot-high fence, and patrolled by armed guards, it became as sacred as an embassy – a portion of the US within Wales. The soldiers who operated it were all ex-railroad men and, on the occasions that we met each other at the gates of the American compound, our life-styles would be eagerly compared. We listened wide-eyed when such romantic names as Union Pacific, Pittsburgh, and Pennsylvania were spoken. Switches were points – switchmen were shunters – engineers were drivers.

They laughed at out 5700 class panniers – 'little matchboxes', they called them. They laughed even more at our rather ancient ex-Barry Railway tankies, as they groaned and wheezed their way around the dockside. One such loco was number 259. She still carried her original boiler, which had gone gracefully weak with age. The red blow-off mark had been painted over on her pressure gauge and reduced to 150 lb. Her firebox was so thin that it was hard to keep her quiet. The standing joke was that you could fire her with a cigarette lighter. Her regular driver was Harry Tiller, a slightly built man in his mid-sixties, who talked, and talked, and talked! Harry just could not stop talking. From signing on to signing off, everyone in earshot, which always included his unfortunate fireman, suffered his constant drone of everyday experiences. I strongly believe that 259 was constantly blowing off steam to drown Harry's voice – but he was invincible. He just could not be subdued.

I had the misfortune to suffer Harry and his engine on a night shift. 259 was more than lively, having a bunker full of patent fuel eggs, very hot and very smelly, giving out a dark yellow sulphur smoke. Harry was deeply engrossed in relating a most mundane domestic situation, when Bill Guppy, the head shunter, arrived on the footplate. 'Fifteen low loaders for the Storage Sidings, Harry, me boy. Sherman tanks they are. Get that done and it'll be time for grub.'

As we ran around the menacing-looking train with our loco blowing her brains out, it appeared to be not too demanding for the veteran.

The Storage Sidings were about four miles from the dock-side, at the top of a steady curved gradient, which ran through a flat roofed tunnel called the subway. Half-way along the journey was a stop signal, where we phoned the signalman, who would allow us all clearance before starting up through the gradient tunnel. 259 was screaming from her safety-valves as we waited to go. Only Harry's lips could be seen moving in the light of the firebox, as the old loco consumed her tarry fuel with a thick yellow fog from her stack. The start wasn't good. We knew we were going to fail, as she gripped the sanded rails with a slow grinding sound: 150 lb pressure was clearly no match for fifteen Shermans. As we laboriously reached the centre of the tunnel, she stopped. If ever I was asked to describe Dante's Inferno, that would be it!

Thick acrid smoke tearing at the lungs; the engine screaming like a thousand witches from her safety-valves – doubts on the safety of the situation – and a white heat from the firebox stabbing the murky interior of the cab. The shunters had gone, and with both injectors trying feebly to cool the raging boiler, I decided to leave also until the atmosphere was a little more tolerable. As I lowered myself from the footplate, my eyes caught Harry in the glare from the firebox – his lips were still moving as he talked on, unconcerned with this most uncomfortable situation.

Within ten minutes, a 5700 class pannier came behind our train, and I ran back into the tunnel to rejoin Harry. 259 had quietened down, and, as I felt my way through the smoke and steam, I could hear his voice: 'You'd better re-light the gauge glass lamp, mate. I don't know why that should have gone out.' I was pretty sure the cause was lack of oxygen, but not enough to stop Harry talking.

The Americans got to know of the incident and the wisecracks came fast and frequent. 'Say, fellas,' said one engineer sergeant, as he leaned over the side of this loco . . . 'You oughta pension off that ol' 259 kettle and keep her for making tea, seeing as how she's always on the boil!'

A confrontation came about on a dirty, drizzly night three weeks later, as we waited aboard a 5700 class pannier tank, near the entrance of the American compound. A particularly brusque American switchman tapped our loco door with his shunting pole. He was a tobacco-chewer, revealing his black-stained teeth, as he smiled up at me: 'You fancy a little bet with this little matchbox of yours?' My driver and the shunters rushed over to my side of the footplate, as he continued, 'We got us a full train down the yard there an' we reckon this matchbox couldn't pull it, but, just to be fair, our loco will haul it up here.' He paused to eject a stream of tobacco juice over his shoulder. 'If you can move that train, we'll give you two hundred Camel cigarettes.'

'What if we don't?' said our driver.

The sergeant's face broke into another black-toothed grin. 'Why, nothing, man! You just remain a lousy little matchbox.'

In austerity Britain, pride had to take second place, and, when honour was the only stake against 200 cigarettes, we eagerly accepted the wager.

As the American walked back to his train, my driver thrust a ball of cotton-waste into my hands. 'Nip down the dark side, mate. Block up his sandpipes with that, and ram it hard. They got steam sanders on them Yanks.'

I could hear the American fireman stoking up his 0–6–0 Yankee tank as I carried out my foul deed, and, as I got back to our pannier, his engine was blowing off furiously in readiness for her duel.

The tobacco-chewing switchman returned to our footplate to act as starter and umpire, and, as he waved his lamp with a green shutter, the American loco began to slip. Myriads of sparks from her chimney lit up the inky black night sky as she failed to move. The switchman cursed, as he chewed savagely on his tobacco wad. 'What the hell's wrong with the goddamned thing? She didn't like that goddamned train, but, by hell, she pulled it right clean through this goddamned yard!'

My driver, fearing that he might rouse our trick, was quick to respond: 'The incline's just a shade more severe from there on, sergeant. Just a bit too much for her I'd say. If you uncouple, we'll take your train from there.'

The sergeant actually did swallow his wad, causing him to go into a violent coughing attack, as he spasmodically replied, with tears streaming from his eyes, '*You'll* take it from there! Hell, man, you don't stand a cat in hell's chance with this matchbox!'

The driver's eyes lit up: 'What's the betting?'

The American took a bite from a fresh plug of tobacco. 'Double or quits. Four hundred cigarettes or nothing!'

'Done,' said the driver. 'A full head of steam, mate, and sand the rails all the way down to the train.'

As our loco gently eased herself onto the wagons for a snatch start on a generously gritted line, the steam gauge came up to full pressure, and we were ready for the marathon. She took up the slack of the couplings, then came to a complete standstill, as if she were anchored to a granite wall. The lever was right out; the regulator wide open, as she breathed the very last ounce of steam into her valves and pistons. She groaned and grunted, as she seemed to flex her steam chest, like a weightlifter. Any second we expected her wheels to spin like tops, but the sand gripped them.

The American spat repeatedly with joyful anticipation at the thought of our imminent failure, whilst we began to abandon all thoughts of our cigarette victory.

Suddenly the situation became as tense as a real-life drama. We all looked at each other stern-faced, as our loco began to move slowly – so very slowly – while her first exhaust left the chimney like a long tired yawn.

The next exhaust was faster, and soon she took on the old railwayman's chant: '*I think I can – I think I can*' and, as she picked up speed: '*I knew I could – I knew I could*.'

The Americans accepted their defeat in good heart and unaware of our trickery, for I removed the waste at the first chance that night. They were generous to a man, however, and many gifts, mainly foodstuffs, gratefully came our way. Cans of chopped chicken, smoked pork, and red salmon, followed by pineapple chunks with thick cream, all were a welcome change from the habitual cheese sandwiches.

One night, however, it wasn't food. Sidi Barrani, our

grade three shunter tapped the door of our ex-Barry. His title was obviously a nickname, and I never did know his true one. 'I got a heavy box down here,' he said. 'Yankee sergeant give it to me. Reckons it's worth a few bob.' It was extremely heavy for its size of only about a foot cube, but it was built like a battleship, one inch planks and iron bound! Bill Hovard, my driver, normally an easy-going type, intervened as we struggled to manhandle it onto the footplate. 'That's not coming up here, boys, I don't mind a bit of grub, but not that, look how it's done up. We don't know what the hell it is, could be a secret weapon.'

A popular fantasy of that time was 'bullion'. I don't know how it started, but talk of 'Yankee bullion' was rife. Any strong-box van with more than one seal was bullion, and if that was remotely factual then the whole contents of Fort Knox had been transferred to the New Dock Sidings at Barry!

The head shunter arrived at the scene. He had a dominant personality accepting loco-men only as a means to shunt his trains and nothing more. 'What we got here then, Sid?' he said to the grade three. 'Looks a bit valuable like, don't it?' Bill Hovard started to object but he was ignored. 'Let's get up against the stop-block at the East Breakwater and open the blighter up, eh?' The line along the breakwater ran between the Bristol Channel and a semi-dry pond. It was a rough night with force eight gale driving a ferocious high tide against the causeway. Under protest, Bill slowly moved the Old Barry in the direction of the breakwater. It was wartime, of course, and we were thankful for the canvas blackout blinds as the constant salt spray thundered against them. 'That'll do,' said the head shunter. 'Now hand me the coal pick, fireman, an' let's see what the sergeant give us.'

A look of dissent from Bill Hovard prompted me to remain neutral, and I pointed to the pick near the damper. The whole procedure looked decidedly criminal as Bill and myself watched the two shunters attack the box with as much zest as a pair of sixteenth-century buccaneers.

The cab could well have been a cave as the firelight of the Old Barry played on their sweating faces, whilst outside the storm raged on. There was a distinct lurch of the old loco which I took to be caused by one of the tremendous hurricane gusts of wind. We craned forward as the iron bands were removed and Sidi Barrani drove the pick into the wooden sections of the box. Suddenly, with a loud squeak of wrenched nails all was revealed – hundreds of bronze-coloured discs. 'Dollars,' said Sidi. 'Four to the quid, and there's hundreds of the blighters, we're made, we're rich.' The head shunter was studying one of the discs. 'You know what these are? Medals, bloody medals, they give these to any Yank who serves in this country, they're not worth a light!' As if to drown his sorrows from the outside a deluge of sea water suddenly tore away the blackout canvas and soaked him to the skin.

Bill Hovard jumped from his tank top seat as the old engine gave a second lurch to be noticed now by the two shunters. The four of us rushed to the seaboard side of the cab where the salt spray stung our faces with its ferocity, but we could see nothing – it was inky black. Bill pulled his macintosh from the bunker locker. 'Well, I'm not going to move till I have a look around.' His patience had worn thin with the shunters. I decided to go with him as a precaution in such atrocious conditions. With only my short jacket as a protection, I followed him down to the line-side. As we struggled forward to the front end, holding on to the framing to keep our feet, wave after wave of the incoming tide pounded the breakwater with a terrific shudder. I tried to grasp a buffer as my feet suddenly went from under me. I could feel one of the sleepers level with my knees as I stood in a hole in front of the loco, about a foot of ballast had been washed away leaving the sleeper and rail suspended. The two shunters joined us with their lamps. They afforded little light as we inched along the hanging sleepers but they did eventually enable us to assess the storm damage. A stretch of ballast about 10 ft in length had been washed away, leaving that distance of rail and sleepers suspended a foot above the granite bed of the breakwater.

We returned to the footplate for a council of war, and it was war, between Bill Hovard and the head shunter.

'You and your mate can ruddy well get down there and pack that rail right now, storm or no storm,' said my furious driver. Fortunately a break in the clouds allowed the full harvest moon to illuminate the whole area as the two shunters set to work. Ironically it was the moon that caused the exceptionally high tide, and ably aided by a south-westerly force eight. 'That'll teach the blighters a lesson,' said Bill as we watched the shunters drag small concrete sleepers from a convenient stack nearby.

'Try her now,' shouted Sidi. I felt decidedly uncomfortable as Bill opened the regulator a fraction and the old loco edged forward. If either side of the packing collapsed the engine would most certainly roll off the lines and over the side of the breakwater. It was only 10 ft, but the experience seemed to last for a mile, as she lurched and swayed over that short unsteady bridge. The full moon remained to show us that there was no further damage, and the two guilty shunters joined us. 'Now get rid of that lot as soon as you like,' said Bill, pointing to the medals. Sidi began throwing handfuls over the side and when the box was empty the Old Barry hungrily cremated it.

I returned to the spot many years later, and made a thorough search between the giant granite rocks of the breakwater, but time and tide waits for no man, and all that remains is my memory.

13 · Union Loyalties

I suppose that trade unions are part of a natural cycle within our industrial evolution. Of their existence, or otherwise, many will argue. During my railway days they had little significance. Only a handful of staunch members would relive the glorious victory of the eight-hour shift and the guaranteed week in 1919.

Their sceptics would respond with memories of the general strike in 1926. 'What did that get for us, eh? Six months loss of pay, and nothing gained after it was all over.'

Branch meetings were very sparsely attended on Sundays, there were no shop stewards – the only official being the secretary, who was ever present on paydays as we fought to extract our compressed pound notes from the tiny one and a half inch diameter pillbox tin that even in those days had not kept up with inflation.

Annual pay increases were non-existent, and there was great jubilation when the employers awarded a rise out of the goodness of their heart.

Regarding working conditions, the sad demise of the steam loco was welcome for it could never be compatible with the modern trade union way of life.

Can the reader possibly imagine a present-day militant fireman pitting his all against an engine that was determined to break him? Or would we see a driver make his own worsted oil trimmings? Would enginemen sign on half an hour or so before their time, and as for the Rule Book, that would have to be re-written!

On the evening that I returned from Swindon after passing my medical, a young man called on me. An obvious product of working-class academic training, he politely introduced himself. 'Harry Stamp, Secretary of the local NUR Branch. I'm pleased to hear of your success today, and I've called to enrol you in the Union.'

Unaware of any other organisation, I promptly paid my dues and received my card.

The following morning, as I signed on for the first time, a corpulent middle-aged driver met me at the time office. 'You young Brock?' he said in a demanding tone. Ignoring my answer, he thrust a form at me. 'Fill that in.'

'What is it?' I said.

His face reddened. 'Union, son. Your application to join the Union. You gotta be in the Union.'

'But I joined, up at the house last night,' I said.

His eyes narrowed; his mouth hardened, and the veins of his temples protruded through the tight skin. 'You – you haven't joined the NUR? For gawd's sake don't tell me you've joined the NUR! You're a footplateman, lad. An engineman. The cream. You don't want to be classed with porters and platelayers and every other Tom, Dick and Harry.' He stopped to wipe his wet brow with a large red spotted handkerchief before continuing his verbal onslaught. 'There's only one Union for you lad, ASLEF – the Associated Society of Locomotive Engineers and Firemen. Now, tear up that NUR card and start afresh with your proper Union.'

His words had fallen on my deaf ears. I bore no loyalty to any Union in particular, and, rather than forfeit my NUR joining fee, I decided to stay with the porters and platelayers. He stormed from the time office when I gave him my answer and I thought the matter had ended there, but I soon discovered that an NUR engineman was very much in the minority.

Union Loyalties

During my very first firing turns on dock pilots, drivers would pose the question quite early in the shift: 'What union you in, nipper?'

If, by a ten-to-one chance, he was NUR, my day would be a pleasant one, but, if he was a strong disciple of ASLEF, the shift would be extremely miserable. But despite the weakness in footplate membership, the only militant action that I recall was carried out by the NUR.

Every shed operated on a link system, and each link was a work style based on seniority. The lowest links, for instance, would be shunter work, whilst the top link would be passenger duty. About half-way up the ladder was the spare link, where the engine crew's duties could vary from relief work to shed engine routine, and that was where the trouble started. A spare driver could only be asked to move and prepare engines, but a fireman could be put cleaning ashpans and filling drams on the coaling stage, if the shed was short of labour.

When I reached the spare link, I seemed to spend much more time than my colleagues filling bunkers and raking ashpans. Each of the three shift foremen homed in on me like a pigeon whenever I had a spare moment in the mess-room, and I was constantly chosen, from as many as ten firemen, to carry out the most unsavoury of shed work.

It only came to light one day when the foreman selected two of us for the coal-stage. His words, as he left the mess-room, were most certainly food for thought. 'Off you go, lads,' he said. 'Horses for courses, and jobs for Unions.'

'You don't happen to be NUR by any chance?', I said to my fellow sufferer. He smiled back. We were well aware that the three foremen were ASLEF members, and it was no coincidence that all their victims for dirty shed work were NUR members.

The Rule Book solved it. That wonderful book of words. That Bible of the railway, that causes chaos if adhered to!

Joe Evans, an NUR dissident, brought it to work one evening. In the first few pages it stated that all servants of the Company should present themselves in a clean and proper manner. One shift on the coal stage would most certainly obliterate the pristine blue of Monday's overalls. The words of wisdom went on further to say that we should protect our health in order to abide by footplate safety regulations.

We immediately thought of the sharp, hot grit of the ashpans we raked. That was definitely not compatible to the high standard of eyesight required by the Company.

For the first time on record a Sunday strike meeting was called for at the union-owned dance hall of the local NUR branch. It was well attended by the younger firemen, but most older men, including all of the drivers, branded us as a band of trouble-making young hot-heads. Driver Tim Cannon, our Secretary, presented an ultimatum to the Company that under threat of strike action we would no longer rake ashpans or carry out coal-stage work. It was a partial victory on our part, for ashpan cleaning, except in an emergency, was abolished, but coal-stage work was to continue without the actual tipping of skips into the locos. So it appeared that the management in their wisdom ignored the dirty job part of our plea, but conceded that our eyesight should be protected.

Their embargo on skip tipping was one we had missed, but the management had decided that the action could cause ruptures. For a while the NUR locomen went about cock-a-hoop over the victory, whilst ASLEF men remained quite sullen, for it should have been a victory for ASLEF who obviously claimed to be the only union for locomen.

The ferocious winter of 1947 must have been the greatest force against the unions. All job routines were ignored, as crews of locomen cleaned frozen points, and many other unusual duties, during one of the worst blizzards this country has ever experienced.

NUR and ASLEF worked hand in hand to keep the lines open, and a common sight would be four locomen to an engine. Two drivers would be in the cab, whilst their firemen would be outside by the smoke box directing steam lances onto frozen points.

About this period a large construction project was taking place in Mid-Wales, and the cement for the contract was provided by the Aberthaw factory in the Vale of Glamorgan. As I remember the requirements called for two trains a day comprising fifteen hopper wagons headed by a '57' pannier tank with vacuum brake. These engines had been brought in from other depots as the trains were vacuum throughout and Barry Shed only possessed 5700 class panniers with steam brakes in the 6700 number series.

My driver on this train during the 'Ice Age', as it was called, was Tom Hayes, an ardent member of ASLEF who often ragged me over my NUR membership. I was never sure if he meant it or not, but he had the distinct attitude of making me feel extremely uncomfortable whenever we worked together. He made a few caustic but humorous remarks when we signed on, and little more was said than was necessary during the long haul to Dowlais.

The weather, though bitterly cold, remained clear of any snowfalls, and, after a short break at the colliery, we coupled up to our homeward train. The vapour from the guard's mouth belched out like a snorting bull, as he shouted up our load from the frozen ground below, and, as a few small flurries of snow swirled menacingly around the cab, we pulled away.

Dusk came in early with the aid of an ominous black sky, and soon we were in the midst of a heavy snowstorm. Luckily, we were travelling boiler-first, for the engine was minus her side-sheets, an absolute necessity when working bunker-first in bad weather. Our train faced the blizzard mile after mile with only the ghosts of signals. It was futile to look out into a hostile world of blinding, driving whiteness. The footplate became a warm, cosy sanctuary, as I sat on my seat between firing to face Tom. His face broke into a slight smile as we steamed on. His contempt of my Union was waning in the emergency of the moment, and he offered me his tea bottle. 'Here, have a swig, mate! I got a clean bill of health.'

As I took a welcome mouthful of the sweet, tepid liquid, I was suddenly thrown against the boiler, as the loco came to a stop.

Tom closed the regulator, as she began to slip as if she were trying to push against a granite cliff face. There was a further thrust from behind, as the train slackened its couplings to put its full weight against us. 'What the hell have we hit?' said Tom.

Taking the boiler gauge glass lamp, I began to descend the steps into snow, deep snow, at least 4 ft of it. With a change of mind, I swung myself out of the cab and onto the tank top to walk forward. We were firmly embedded in a snow drift as high as our chimney! We had entered a cutting, possibly 10 ft deep, and packed to the top with snow.

Tom knew the spot quite well. 'It's about half a mile long,' he said. 'Lord knows when they'll dig us out! Better settle down for a long stay.'

I began to cool the fire down by closing the dampers on the engine but the big fire wouldn't die easy, and she blew off relentlessly for an hour or so. I was aware of the low level of our water tank, as the red pointer hovered near the bottom of the gauge. The water column was just at the end of the cutting for I had noticed it on the up journey. Only a half-mile away, but it might as well be in Siberia!

Tom looked back along the train. 'No signs of the guard, mate. Still, there's not much any of us can do. No sense in risking a broken ankle. We're safely locked in this section by both boxes, so we'll wait till someone calls to see us.'

I began to search my food bag for any uneaten sandwiches, as my mind turned the simple affair into a ship's lifeboat drama! Tom watched my futile attempt in searching the empty bag. 'Here,' he said, offering me his small sandwich tin, 'there's one left in there.'

I reluctantly refused out of politeness.

He pushed the tin into my hands. 'Go on, eat it! I've got a few beef cubes. A couple of those will do me. I don't eat a lot these days, 'sides, it'll do my ulcers good to starve a bit.'

The glorious memory of that single cheese sandwich, and a steaming hot cup of beef extract made from boiler water, remains with me to this day.

Soon I fell into a contented, deep sleep. The bright dawn of the vivid white countryside, combined with a shoulder-shake from Tom, awoke me stiffly from my slumbers.

'The water's gone, mate! Injector just blew out, and we got about half a glass in the boiler.'

I looked back at him with helpless, glazed eys. What could we do about it? Only drop fire and save fusing a lead plug . . . Then we'd freeze to death, or at least be extremely uncomfortable on an ice-cold loco miles from anywhere.

As we drank the last of the beef cube mixture an hour or so later, the water began to play hide and seek in the bottom nut of the gauge glass.

'That's it then, mate,' said Tom. 'Let's drop fire. Can't risk it any longer. You start an' we'll take turns at it.'

I realised that he was being considerate to my youthful experience, for the clean fire on top would be quite easy compared with the hard clinker below.

My first shovelful of red-hot coal vanished into the deep snow with a hissing explosion and a cloud of steam. Suddenly it came to me. Snow! That should melt in the warmth of the tanks. 'How about it, Tom? I said. 'I suppose we'll need a hell of a lot of it, but we've got nothing else to do.'

He agreed, and for the next two hours we took turns at

packing the snow into our tanks.

Within twenty minutes, we put one injector on and also managed to maintain the meagre level on the indicator for half an hour. After two hours, we had a nearly full gauge-glass but precious little water in the tank. That snow just vanished into nothing once it melted.

Returning to the footplate, we both experienced that feeling of glowing health after working, or playing, in snow.

Tom wiped his brow, as we sat down to face each other. 'If we keep this up, mate, we'll dig ourselves out, and I wouldn't say I'd mind if we had a bit to eat and a brew of tea.'

As the morning wore on, we chatted amiably on many subjects and never once did the matter of Unions ever raise its head. We were true enginemen working in unison.

Suddenly came half a dozen shrill blasts of a loco whistle ahead of us. The snow-plough was working its way up the cutting. An hour later, a permanent-way inspector appeared on top of the embankment to say we would be free to proceed with caution.

Twenty-four hours after signing on, we returned to Barry Shed, very tired, but lifelong friends.

Finally on the subject of Unions, a footplateman might do worse than follow the ways of driver C.A. Makepeace, who left the services of the Brecon & Merthyr Railway in 1889 to become one of the Barry Railway Company's first drivers. He was a very active member of both the NUR and ASLEF, and to top it all a staunch Conservative. I very much doubt that membership of any two of these three organisations would be remotely compatible these days, but he obviously lived up to his name of Makepeace.

14 · A Fishy Story

Barry Island, the Blackpool of South Wales, could attract as many as 100,000 trippers in one day. Barry Shed on such days could completely lose its character when a few dozen foreign locos, mainly from the Midlands, lodged there for the day while their crews also took to the seaside. The extra trains required for the influx of 'Valley' people was usually shared between Barry and Treherbert 'specials'.

I had a distinct premonition of impending disaster as I trudged the ash path toward Barry Shed on a sweltering August Monday mid-morning to work a special. Despite the late signing-on time I still nursed a delicate head in memory of a long night out at the Railway Club not many hours before.

Ted Walker, the time clerk, threw me my engine keys. 'You don't look man enough for her this morning, Brocky boy.'

I grasped the one-inch nut with the key and number tag threaded through it. 345, a Taffy, and a fussy old bitch at that! Why, oh why, couldn't I have a '56' on a day such as this. I considered a sudden attack of flu, or appendicitis, but discretion won the day, and furthermore I needed the money.

She was second off in the shed, and the cleaners were just leaving her, but my luck was beginning to desert me as they rushed for the newly opened canteen. Normally they would clamour to carry out most of the prep work.

My driver Charlie Burley arrived, and I drove her to the outside pit for oiling. She had a nice back fire of Welsh large, and the steam pressure gauge showed 100 lb. 'Not so bad,' I thought. 'Might even get to enjoy it after I've worked off the ravages of the night before.' And then it happened. As I opened the bunker slide an evil-smelling black sludge oozed onto the footplate like a scene from a science fiction film.

'That's it!' I said aloud to myself. 'That's the bloody last straw. 345, with a bunker of the worst coal I've ever come across.'

It didn't miss the dry wit of my driver as he came on the footplate. 'What have you got there, mate? Mortar? You going bricklaying?'

I was in no mood for jokes. There were a small minority of shedmen who bore a grudge against enginemen, and I knew of one particular character on the coal stage who would delight in sabotaging the temperamental 345. With a bunker of best Welsh large she would sing her way all through the valleys, but anything less meant the kiss of death for the best of firemen.

The coal-stage worker hadn't been cunning enough when he'd sandwiched the slurry between the best large, for the steam-raiser had used up all the bottom layer to expose the inferior coal, if it could be classed as coal.

Mr Chidgy, the foreman, had just a few driver friends, but he was certainly no great admirer of firemen. He looked up from beneath the brim of his trilby when I burst into his office. 'I can't take that loco out, foreman, she's got a bunker of rubbish. It'll put the fire out instead of making steam.'

His face remained expressionless. 'You haven't got a lot of say in the matter, mister. In any case, drivers decide on a loco.'

In a brief moment of compassion he repented slightly. 'It's the nationalisation, that's the cause of it. The government haven't got the money the old Western had. The days of good coal have finished.'

I knew he was wrong. The little steam dredger aptly named the 'Graball', had for years cleared spilt coal from around the bases of the coal hoists of Cardiff and Barry docks, but her harvest had never found its way into the bunker of a passenger engine before. British Railways were not to blame for my misfortune.

I insisted with the iron-hard foreman. 'It's the coal stage,' I said. 'To put that stuff into a Taffy is criminal, even if she wasn't booked on a passenger.'

Mr Chidgy rose to his feet. 'That's enough now, mister – if you're not going to work her then say so, and we'll see what the divisional super will say about it.'

He had the best of me, as a visit to the great man invariably resulted in a week's suspension. A cynical laugh came from beneath his trilby as I began to leave the room. 'Don't worry, mister, if you do stick for steam out in the wild you won't starve, will you, with all that shell fish.'

Threequarters of an hour later we coupled up to our train of six coaches to await the instructions of traffic control as the heat of the day began to intensify.

With both dampers open for half an hour the steam pressure gauge didn't budge from 195, whilst the dead fire would most certainly have welcomed some harsh treatment from the blast pipe. At long last our guard shouted from below. 'Empty stock for Porth, driver – soon as you get the board.'

It was another half an hour before the signalman could allow us to cross Barry Junction, as train after train of day-trippers headed for Barry Island. The trip to Porth went reasonably well as far as the engine was concerned by keeping a small back fire, and lower still at the front of the box, but very bright.

The nauseating smell of the decaying mussels combined

with the oppressive heat of the day did little to relieve my thick head, but on taking water at Porth I decided that a welcome accident would take place. By bending the leather water column hose in the mouth of the tank top I created a gorgeous cold shower that soaked me thoroughly to the skin. The incident caused a good laugh all round, to my driver, the station staff and several passengers, but the cooling down was well and truly worth the humiliation. The trip back to Barry Island went quite well, but of course it was all downhill gradients. We fully expected to wait for orders at the Island yard, but the valley people were descending upon the 'trippers' paradise' in their thousands, and as soon as we took water, it was empty stock once again to Porth.

The fire was becoming decidedly dirty, not a clinker, which could always be broken with the pricker, if caught early enough, but a dense, lifeless, black dust. It was obviously fine enough to go through the fire bars, so I assumed that the ashpan was full despite both dampers being open.

We ran into Porth with 150 on the steam pressure gauge, and a half-glass of water, and me praying for a drop pit to clean the solid ashpan, but that was not to be, for Porth did not possess one. By the time we had run round, the train was packed and 345 was most certainly not ready for departure.

More as a matter of morbid interest I lay on my tummy in the four foot to view the lifeless ashpan. Why the cursed ash did not blow away with both dampers open to this day is still a mystery to me. Suddenly I heard the metallic clang of the starter signal dropping. Rushing back to the footplate I noticed the steam gauge just on the blowing-off mark, but I could hardly feel elated, for we were still stationary, with dampers open, and the firebox doors closed, enough to tear the boiler apart on a healthy engine!

As my driver blew off the brakes, I looked back for the guard, and then the horrible truth began to dawn on me as the bewildered station staff also stared at the rear of the train. The Rheola Arms just outside the station was open, and our guard had more than a passing affection for his pint.

The head porter tapped the footplate door. 'Your guard haven't come back, driver, an' I got to get you away.'

Charlie Burley, a true servant of the Company, toyed with his watch nervously. He would have preferred to work to the Rule Book and refuse to move until he was flagged on by his guard, but there existed a strong code of honour amongst railwaymen.

'Why don't you nip up to the pub and get him?' said Charlie.

'Haven't got time,' said the worried-looking foreman. 'The regular service train from Merthyr is due in any time now.'

The shrill blast of a steam whistle solved any doubts of pulling away, guard or no guard. The regular service train was held up outside the station, and we were five minutes late, with six sardine-packed coaches of day-trippers eager for the delights of Barry Island.

With a reluctant driver, we started away, and, to mimic his feelings, the Taffy at last began to oject to her diet of mussels and small coal.

Despite a downhill gradient most of the way, our destination could not come soon enough. With the water bubbling in the nut, and my cap covering the steam gauge, we ran into Barry Island station platform, with the blower hard on, fifteen minutes late.

With both injectors on at full bore, to fill the sadly depleted boiler, we ran around to head the train in readiness for yet another trip to Porth.

With our starter signal at the 'off', the busy station staff, who were anxious to see us away, had not noticed the absence of our guard, who was happily consuming pints some thirty miles away! I conjured up envious pictures of him, as I fastened the water column chain to the tank lid. Quite a crowd of children and parents had gathered around the water cock as I controlled it while my driver carried out his oiling. Now the Rule Book stated that the fireman and driver should carry out the operation of taking water, but most railway rules had to be broken or the system would grind to a halt – hence the reason that 'working to rule' is a very effective industrial action.

That particular day was most certainly not ripe for rule-breaking, for I noticed that a middle-aged military type of chap, with a thin waxed moustache, eyed me sternly from the rear of my audience. Despite the intense heat of the day, he wore a rather dirty gaberdine raincoat and a battered brown trilby hat. Suddenly he pointed at the leather water hose. 'Unfasten that chain at once! You know it's a two-man job.'

I was convinced he was some kind of an over-zealous railway enthusiast, and I laughed back at him. His face reddened, as he spat his words at me in army fashion. 'Didn't you hear me, man? If you and the driver do not take water in the correct manner, I shall have you suspended!'

I was in no fit mood to humour him any longer. 'Oh, belt up!' I said. 'Who the hell are you anyway?'

He pushed his way through the crowd towards me. 'I am Major Trippet, Divisional Superintendent of the Line.'

I had heard of him, a bowler-hatted ex-army man, newly appointed to the managerial staff of British Railways. But I had gone too far, and I just couldn't retract what I had said in front of all those people.

'Now you just listen to me,' I said, as my blackened sweat-stained face nearly touched his. 'This engine needs water and oiling, and she's running late because I can't make her steam on mortar and mussels. If you can do better, have a go, because I'm going!'

At that moment Driver Burley appeared around the front

end of the loco, holding his fingers to his lips. His face portrayed a mixture of fear and utter disbelief that I, a humble fireman, would dare to talk so aggressively to such a superior.

But my verbal onslaught had taken effect. The great man began to capitulate, perhaps at the thought of handling a shovel for possibly the first time in his life. He held his hands in the fashion of a vicar. 'Don't be too hasty, lad. I can see you have problems. Pull away as soon as you can and we'll forget all about it.'

My audience, fully absorbed in the drama, edged me on. 'Go on butty! You tell him,' said a blue-scarred miner.

The tanks, of course, continued to fill, as the Superintendent stood perilously near to the inlet. Suddenly, he was drenched to the skin with the deluge of the overflow. The crowd roared with laughter, as the miner once more became their spokesman. 'You asked for that, mate! You shouldn't have interfered.'

The situation was a true-life comedy – station staff frantically trying to start a train that had no guard; a very wet Superintendent trying to persuade a reluctant fireman; and a hostile crowd of passengers and onlookers adding their force to the affray.

The miracle was that with all the pandemonium, no one missed the guard. When we eventually pulled out of Barry Island with promises of no disciplinary action from the pathetic sodden figure of the Divisional Superintendent, we gave a sigh of relief.

The same worried porter met us at Porth. 'You missed him again then, did you?'

'Missed who?' said my driver.

'Your guard, your guard,' said the porter. 'He came back from the Rheola just as your tail-lamp left the platform. He worked it out that, if he caught the service train that followed you in, he would get off at Barry and get the signalman to stop you on your way back from the Island.'

I remembered that we passed the service train as we pulled out of Barry Island, and, amongst the seething mass of faces of the happy day-trippers that rushed past between the narrow space of two trains, I might have seen the face of our guard.

We did eventually meet up with our elusive crew-member when we returned to Barry Island, and the Taffy was done for. It was intended that we were to be relieved, but Charlie informed the Control of our problems and thankfully we were instructed to go on the shed. I could never blame the engine as I passed her on the drop pit, but I couldn't conceal a deep-rooted grudge against the person who had caused it all.

Some months later I was working a berthing shift at the coal stage when I was aware that my suspect was one of the coalers. By an uncanny coincidence 345 came through, and as I berthed her, he came out to see what loco he had to coal. The first dram came out, beautiful shining large. Our eyes met as he tipped load after load of perfect coal into the temperamental Taffy, but finally he gave himself away. 'Bit better than mussels Brocky!'

Although I have gone through life in a fairly light-hearted manner, the day of the Porth special was beyond a joke, and revenge began to cloud my mind. The first chance came when the Bottom End shunters were restocking the coal stage. I noticed that the contents of two 20-tonners could not be seen from ground-level. One contained Ocean colliery large, the other patent fuel eggs. I changed the labels.

From my hide-away outside the stage I watched my victim toss the coin with one of his colleagues for choice of wagons. The winner would obviously take the fuel eggs, for they would load themselves into the first half-dozen drams, but you had to control them by using the wagon side door as a hopper. My victim lost the toss. My evens gamble had paid off, and like a lamb to the slaughter he picked up a sledge hammer and walked over to the wrongly labelled wagon of fuel eggs. Now, you could hit the latch pins on a wagon of large, and the door would drop to reveal a solid wall of coal, but eggs would run like sand. He didn't even bother to place two drams at the ready. With a bad-tempered swing of his hammer one pin sprung to swing on its chain. My heartbeat increased as the hammer sped towards the second pin. *Whoosh*! Ten tons of fuel eggs spewed over the coal-stage floor burying my victim to his knees. Of course, he had to clear it all up and coal his quota of locos.

At the time I felt that honour was done, but during break-time another fiendish idea crossed my mind, as my victim chopped his giant bundle of firewood sticks. The planks of timber came from the wagon repair shops nearby. There was a communal hand-saw and axe under the coal stage where each man would make a wheel of sticks, by hammering the firewood into a wire band. Some men took quite a pride in their firewood wheel, and my victim was undisputed champion. As the break-time finished I eagerly watched him take the stairs to the coal stage above and I began my dirty deed on his wheel.

It was many weeks before we met again outside the roster room, and he did look quite menacing as he approached, but the joke was on me. 'How did you know I needed half a dozen bricks then, Brocky boy?' Then I recalled how I had taken well over an hour to hide six bricks in the centre of his firewood wheel which he really needed for some job around the house.

15 · End of the Line

Footplatemen had always been regarded as the aristocrats of the working-class, right up to the 1950s when their standards began to decline. In pre-war days, I well remember my father, who was a carpenter, refer to my engine driver uncle as 'a five-pound-a-week man', whilst his own pay packet rarely exceeded three pounds.

Single young firemen were eagerly accepted into the homes of their girlfriends by parents who wished to secure a sound marital future for their daughters. But the early post-war years brought a period of unrest to the young men of the industry. The barrack-room philosophy of demobbed servicemen began to penetrate their outlook. 'I'm going to better myself' – 'I'm fed up with shift work' – 'I'm looking for a clean clothes job' – all were common phrases.

I was no exception, having been forced into the industry in the first place. Twenty-six different shifts in a year were never conducive to my way of life, and the odd times, such as 12.05 a.m. or 2.55 a.m. for signing on, were completely alien to my untidy mind.

There was, in fact, a form of revolution taking place on the part of the younger men, and much of the old iron discipline of the Great Western was being ignored as British Railways took over, but the rot had set in, and a mild atmosphere of discontent most certainly existed. A half-dozen or so of the young firemen of Barry Shed joined the police, whilst others went to the new chemical factory within the town.

At the age of twenty-four, I still had futile dreams of being a sea-going engineer, rewarded only by fantasies, as I gazed enviously at the engineer of the local paddle-steamer during day-trips to Weston-super-Mare.

It was during one evening at the local dance hall that I met the then girl of my dreams, and there was little doubt of her being a cracker, for there was a constant stampede of local beaux each time a dance began. I managed to hold her attention for most of the evening and, as we got into deeper conversation, I discovered that she was a train inquiry receptionist at Newport station, and also the Carnival Queen of that town.

'What do you do for a living?' she said, as we stood outside Bindles Ballroom, near the romantic setting of a moonlit Bristol Channel.

'Oh, me!' I stammered, as I tried to think of any profession other than a railway fireman. 'Why, I'm a salesman. A travelling salesman. You know – a commercial traveller.'

Her eyes sparkled. 'Oh, that must be interesting! Travelling all about the country. You have a car then?'

This, of course, was the ultimate status-symbol of those days.

'Er – no,' I stammered again as I fought for more lies. Car? I couldn't even drive one! 'No, I haven't a car. I'm on trial for a year. I'll have a car after that, if I'm successful. I use public transport at the moment.'

Her sparkle vanished slightly. Newport was twenty-four miles away.

But despite the lack of transport, the courtship developed. It became difficult to explain my absence when unsocial shifts came between us. I even offered other firemen cash inducements to change with me.

I began to wear gloves to protect my hands and give them

a salesman's appearance. But gloves and the shovel never were compatible to me, and I abandoned the idea in favour of more expensive hand barrier cream, after I had thrown the shovel into the firebox of a branch passenger loco. In fact, my hands were in a worse state after firing by hand until we reached Barry station, where I suffered the humiliation of being presented with a shovel by a giggling engine cleaner.

I managed to keep up the façade of my Jekyll-and-Hyde existence for three months, when, on one extremely hot midsummer day, I worked a mixed goods to Severn Tunnel Junction, where we were relieved. We were hoping for a trip back on the 'Cushions', but that was not to be. A down mixed freight, with a Tyseley crew, wanted relief, and we were the victims of the 'County' class tender engine heading it.

A very tired-looking fireman met me on the footplate. 'Ever fired one of these before, mate?'

They were one of the new post-war additions to the Great Western fleet, and completely unknown in South Wales. 'Apart from a dirty fire, the ruddy thing's got two blast pipes. Can't keep nothing in the box – only clinker – and she's got plenty of that! He swung his bag over his shoulder and was gone.

I didn't fear much as we pulled away. Cardiff was only about thirty-five miles, and all on the flat, but did she bark! The coal was virtually sucked from the shovel, as my driver tried to get her under way, and the little white feather soon vanished from the safety-valve.

Half-way between the Tunnel and Newport, we were in a desperate situation, as the driver and myself fought to maintain the boiler and 160 lb of steam. I had discarded my overall coat, shirt and vest – down to my bare back – as I tried to break some of the rock-hard clinker with the chisel bar. The terrible disgrace of 'sticking for steam' carried no excuse. The action was so ultimate, so terminal! Every other engineman would say, 'Oh, it could happen to anyone'. Yet he portrayed a feeling of scorn upon you for letting it happen since it was unforgivable in the world of an engineman.

I hadn't noticed the approach to Newport station, as I continued my battle against the stubborn loco. My eyes, red-rimmed from sweat-burn, peered along the up and down main that ran through the centre of the station. The starter was at danger. I prayed that it would remain so for a long time, as my driver closed the regulator.

With the welcome sound of steamless motion, and both injectors filling a boiler dying of thirst, I let myself sag outside the cab in an effort to cool down.

'You look a bit hot and bothered for a salesman.'

We had come to a stop, right outside the Inquiry Office, where my girlfriend was using her break to sun herself on a platform seat.

That scene signalled the end of my railway career. Not so much that she saw me as a very exhausted coal-black fireman – more so the fact that I had lived a lie with her for three months.

That weekend an advert appeared in our local paper for an insurance agent. The basic salary was only two-thirds of my fireman's wages, and there was a commission on collections and new business, but all the insecurity of my future was veneered over with dreams of becoming a 'nine-till-five man'. From then on I would abandon my overalls for a smart suit. No immediate grandeur of a company car admittedly, but my bicycle, with a brief-case strapped to the cross-bar, would suffice for that – my first launch into the world of commerce.

My letter of notice resulted in an interview with the 'Shed Master', Mr Doughty. The action of reaching the inner sanctum of this great man was known as 'going on the mat'.

End of the Line

I met him at his office door, as he arrived for my early appointment.

Despite the intense heat of the huge office coal fire, he remained as cool as his character, beneath the size-too-small black bowler, and the overlength black cloth macintosh of the Great Western Railway Company. His brown boots squeaked in chorus, as he eased himself up and down on his toes. 'No one in their right mind leaves the Western, lad – except in a box, that is! You sure you haven't had a bump on the head? Do you fancy a trip up to Swindon to see the Company doctor? P'r'aps a spell off the mainline might put you right. Come on lad! Think it over.'

Even as he spoke I had misgivings about my decision to leave. He was right, at least until the recent exodus of firemen to other jobs. The unheard-of was taking place. The 'cream' of the railways was actually leaving the job! Despite the trials and tribulations of a decade of unearthly hours worked in sweat, coal and grease, I suddenly wondered how life could exist without it, but the lure of a 'clean clothes' job was too strong.

'I'm sorry, Mr Doughty,' I said, 'my mind's made up.' Later that week in 1953, as I walked along the ash-path for the last time, I felt a deep pang of regret well up.

I've retained 'clean clothes' jobs ever since, and I never married the girl who forced me into one.

Epilogue

Not many years after the passing of steam, during a Christmas shopping ritual, I took my seven-year-old son to see the festive wonders of our local department store. Like magnets we were drawn to the model railway layout, where a GWR 'Hall' class raced around the perimeter with a full set of passenger stock. Having paid our respects to Santa Claus, who sat near the layout, my son wandered through the toy department, while I stayed glued to the little world of make-believe steam dream of bygone days.

Suddenly a voice said 'Reckon you could keep steam on that little lot?'

I looked around, expecting to see one of the many enginemen that I meet in the town from time to time, but I was the only adult spectator, or so I thought, until Santa Claus spoke again.

'Bet you don't know who I am underneath this garb?'

He spoke in a quiet tone, obviously cautious of maintaining his image before his young visitors. With only the sparkle of his eyes as a clue, I tried to place his voice when he spoke again through his white beard.

'Dick Stonehewer, I am. Been retired four years. Couldn't stick those ruddy diesels. Packed in four years early.' He whispered across to the young lady who attended the layout. 'No diesels on this company is there, love? I don't allow them, do I?' He pointed to the little 5700 class pannier, as she picked up speed with her train of coal, while descending an incline. 'You know, Brocky, I must have watched that little blighter coming down that bank a thousand times, and I always think of Joe Rushton and the Big Hill.'

I began to smile, as the legendary story came to life.

The Big Hill was a long incline from a South Wales colliery. It was notorious for its runaways, being well guarded against the mainline at its junction with a set of catch points and a long sand drag.

Driver Joe Rushton nursed his '56' with her full train of coal over the brow of the hill, as the brakeman awaited his signal.

Fireman Arthur Daniels sat back on his seat contentedly facing a full firebox and boiler in readiness for the long haul to Barry once they joined the mainline.

But half-way down the hill the two enginemen looked at each other nervously. The train was gaining speed. Joe Rushton applied the engine brake slowly, as his fireman pulled the sand-pipe lever. The smell of crushed sand prevailed, as the locked engine wheels took on a skid. After several brake applications, the locomen realised that they were not in control of the train as she continued to gain momentum.

Looking ahead, the board was at danger, whilst beneath it the hungry jaws of the catch points remained open to derail the whole train.

Driver Rushton pointed towards the fireman's set of steps. 'Get on the bottom step, mate, and pick a good place to jump off. I'll do the same. I'm not going to risk getting mangled up in this lot when she hits the catch points.'

Fireman Daniels was the first to jump when he saw a clear section of grass verge. He hadn't bargained, however, on the speed of the train, and, though he landed squarely on his feet, he couldn't stop running – right through the open doorway of a platelayers' cabin! It was break-time, and several cans of tea were spilt in all directions before an audience of startled platelayers, as Arthur hurtled along the table like a skittle ball.

Joe Rushton had better luck in making a soft landing the other side of the track, but the highlight of the undying legend was yet to come.

Tom Aikens, the guard, who was rather short-sighted, sat out on the veranda of his van enjoying the views of the passing countryside, unaware that his train was out of control, and worse still, crewless. As the van passed two wildly waving locomen, Tom rushed to open his doors, thinking they required a lift. It was some time before the running men managed to board the van to a startled guard, who said 'What the hell are you two doing here?' Then his face whitened. 'Well, who the hell's up front?'

Miraculously, the train began to slow down as the gradient levelled out, and eventually came to a stop. The three railwaymen slowly left the van on shaky legs to walk the length of the train. Joe Rushton was the first to reach the loco, as she blew off steam wildly. A broad smile came over his face as he came to the front end, where her leading coupled wheels were just about two feet from the gaping trap points.

When I came out of my dream of the past, Santa Claus had just dispensed with a further batch of visitors and was free to talk again. 'You'd never believe the memories that I conjure up by looking at that ruddy little train set all day long. Look at that 'Hall' class going at it! Worked those many a time out of Hereford for Ludlow, Salop and the North.' His eyes misted. 'You know, Brocky, I look forward to this job every year, just to watch this train set! Sometimes I swear I can smell the smoke and oil. And I often picture the characters who worked 'em – them and their fancy nicknames –'

Once more my mind raced back to the past, as I left Dick to contend with the Christmas wishes of a little boy and girl, who stared with starry eyes at the disguised old engine driver.

'Nicknames'!

Everyone seemed to carry a nickname on the railway, and usually for a good reason, as in the case of 'Darkie Dickins', the driver. He always forgot to replace the cylinder lubricator filler cap. Consequently, when it was turned on, a fountain of hot, black oil would splatter the cab roof and rain down on him.

'Bunker Nut' was a shunter who once asked a fireman for some coal for his cabin fire. The enthusiastic fireman was quick to respond, and 'Bunker Nut' was christened with a lump of best large.

The longest nickname must go to a foreman shunter at Cadoxton Yard, who had a habit of saying 'We had a hell of a day last night', and, to the end of his working days, that complete sentence was his nickname, despite its length.

A shedman, who constantly cadged cigarettes, was known as 'Charlie Fag'.

Any engineman from Hereford was automatically branded as a 'Whiteface', after the famous cattle of that county.

'Boogie Woogie Germon', an engine cleaner and grandson of the famous Dick Germon whom I have mentioned earlier, acquired his name in a rather pagan manner. He was call-boy on the West End rota, and, during the lapse of an hour between calls in the small hours of the morning, he decided to take a nap inside a local church. As the week wore on, and possessing a musical bent, he was drawn towards the fine organ of the church. The climax came on the last day of the week, when he could no longer control his desire to play the instrument.

Soon the walls of the magnificent church vibrated with the then popular music of Glenn Miller, only to be interrupted by the local bobby on his beat, who must have doubted his sanity and his hearing.

No one seemed to know the source of engine driver 'Camel' Tom's peculiar nickname, albeit that he had a pronounced hump on his shoulders, or that he constantly spoke of his service in the Camel Corps during the First World War.

'Tom the Rag' emerges again in a tale of derailment one nasty winter's night. Tom's loco came off all wheels on a crossing near Barry station. It was close to midnight that the breakdown gang of the running shed arrived. The weather was foul, with driving drizzle rain, and the tempers of all concerned were rather delicate. It was the first breakdown call-out for the new foreman fitter, who had recently transferred from Canton shed.

Tom spoke to him from the comfortable warmth of the footplate: 'Taking your time putting her on, ain't you?'

The rain-sodden foreman exploded. 'Come down yer an' say that, mate! I got enough troubles without you taking me for a fool.'

Tom's temper also flared, as he jumped down into the four foot and began to fight the unfortunate foreman, shouting 'It's you that's taking me for a fool. I always talk like this. I was born like it. I can't help it.'

'Well, neither can I,' said the foreman-fitter, as he sparred up to Tom. 'I was born like it too.'

Suddenly the two men stopped fighting to break into uncontrollable laughter before their rain-soaked audience in the half-light of paraffin storm lamps. The joke was that they both had nasal restriction and spoke with the effect of having clothes pegs pinched to their noses.

I finally walked out of the store, leaving Dick Stonehewer to his young visitors – his train set – and his dreams.

Many years later when I decided to write this book, I called on Dick Stonehewer to see what I could sift from his memory bank. I took with me a few large bottles of beer to loosen him up.

For two hours or more he carried on in precise verbatim his memories of working from Hereford shed. I didn't have the heart to stop him, but he wasn't telling me what I wanted. I was seeking the more unusual facets of a railwayman's life. 'Halls' and 'Manors' running well or struggling against the odds were so much a part of Great Western life, but that was the normal way of things and not the material for me.

As the evening sun went down and the room darkened, Dick continued to pound the rails along the Welsh and English border, Leominster, Salop, Hereford and Abergavenny.

Suddenly, he realised that I was not completely interested. 'Tell you what though, my old mate, why don't you put a letter in the local paper, there must be dozens of old drivers living in the town who can remember some of the more unusual parts about the job.'

I wasn't desperate for material, but there may have been some gems that I was unaware of.

The letter inviting railwaymen of all denominations to contact me with any unusual stories was completely fruitless with the exception of a phone call from the Secretary of the retired members section of the Railways Staff Association. 'Come to our meeting at three o'clock next Thursday, and you'll have enough to fill two books,' he said. It was a beautiful and innocent con-trick on the part of the Secretary, when I was led like a lamb to the slaughter. I thought I would mingle with the sixty or more members, but that was not to be. I was placed between the Chairman and the Secretary, who addressed the audience.

'I'm sure most of you will remember the gentleman on my right, Mr Derek Brock. He was a fireman about thirty years ago, and he's writing a book about the job, which will be the subject of his talk.'

Apart from the fact that I sadly lack the skills of a public speaker, I was not even prepared for the ordeal. But with most of the book content firmly embossed in my mind, I spoke about it for one hour, when I was given long and loud applause by my appreciative audience.

The Secretary gave me a nudge, 'Mix amongst them now, go on, you've revived their memories, you've got them going.' But they were mute, they wanted to hear my stories. They had nothing to relate, because they didn't consider them to be of importance in years gone by. They had never looked at themselves from the outside as I had. They couldn't possibly absorb the sweeping surge of steam nostalgia that is shared throughout the country, and there was a quiet cynicism for the buffs who support it.

'Most of 'em wouldn't do it if they had to for a living,' was the general comment. Yet I reminded them of their pride in the job, and quoted an instance of signing on before time over a period of forty years which make a conservative total of over 2,000 hours.

'Who would do that these days?' I said.

'Ah! They got more sense nowadays,' said a chorus of ex-drivers. 'We were bloody fools!'

But extremely efficient fools, I thought.

And so, as I neared the end of my book, I had a compulsion for a final pilgrimage down memory lane.

The stroll along the ash-path to the shed was exactly the same as forty years ago, but the canteen is now a motor panel-beaters' shop. Then as I neared the drop pit road for the first time in nearly thirty years, I experienced an uncanny sensation. Silence! Never in its working history was it silent. In varying degrees of intensity it was always alive, right around the clock. The old corrugated iron administration and mess-room block had gone, so had the coal stage. The pretty purple-blossom buddleia plants had taken over where glowing mounds of ashes once burned with acrid fumes. Only one set of the shed doors was open. I peered inside to find a wagon repair shop, which at least had some connection with engines that once lived there.

Outside on the six oiling pits something was blatantly missing, the area was naked. Ah yes! The water columns, nine of them with their leather trunks hanging limp like so many elephants. Even those had characteristics – a soft one was easy to bend into the tank, but if the hole was not lined up true, then it would bend and cause a blockage. If the fireman couldn't turn the cock off fast enough the giant hydraulic sausage would almost explode and drench everyone within range. Conversely, one made of stiff hard leather always objected to entering a tank hole and took the greatest delight in flying out of the tank at the slightest provocation.

I always seemed to do battle with water columns, but a good white fire would dry my sodden trousers in minutes. Suddenly I was drawn to High Level, where the last water column of the area remained. I had seen it from the town many times, but now it was all important to be near it, touch it, and relive memories. On the way I became aware that Barry Locomotive Workshops was also a buddleia plantation. That great building had certainly been unique in its day.

What other small Company such as the Barry Railway could boast its own complete workshop which was actually capable of building a locomotive if desired? It was completely self-sufficient, even to its own sign-writers who shared their skills between the locomotives and the neighbouring passenger coach workshop. The old Barry was sheer perfection in running a railway.

I stopped to scan my eyes across the dock skyline. The last two coal hoists had gone, so had the eleven curved tracks leading to them. I climbed the wooden steps of a shunter's platform, now exposed since the sea of coal that surrounded it had vanished to make it look so unwanted. I reached the site of the double-storey yard cabin. It was gone – replaced by the buddleia. Who, in the wonderful action of nature, tells this plant that a building has been demolished?

On closer inspection only the water tower remained. The swinging arm and leather bag had gone, along with the ladder that we once climbed to take a swim in the tank above. Yes, the dock pilot firemen and some of the younger drivers had been keen swimmers. Undeterred by the murky waters, they took the plunge whenever possible during the summer months. One notable dare-devil once dived from the top of a coal hoist, a distance of about 100 feet. I strolled on to East Dock, and the picture was the same: no coal hoists, sidings, or traffic cabins, not even a water column. Cadoxton Yard is now a large housing estate. It's hard to believe that '56' class 0–6–2 tankies once struggled with their long trains of empties over ground now occupied by well-kept gardens and the lounges of the new houses.

The lines to the valleys via Wenvoe Tunnel have gone – I hope the Royal Family know this – and the coastal branch line to Penarth is no more.

How the old Barry Railway has been devastated, not so much by Doctor Beeching but by the death of King Coal.

I finally returned to the starting point of my reminiscence, Woodham's Yard. Very little had changed, a couple more locos had gone to the good homes of the preservationists, and a few more brave societies have started on the mammoth task of saving their charges from the slaughter house. The elements seemed to have increased their ravages on the remainder, many of which are probably doomed to the furnaces of Port Talbot steelworks along the coast.

As I rounded the buffer beam of a loco the silhouette of a solitary figure headed towards me through the two rows of engines. The rolling gait and slightly bowed legs belonged to a driver whom I hadn't seen in many years. I fought to remember his name, but Jack will have to do.

He rapped his knuckles against the tender of a '28' as we met. 'You can't leave these things alone then, I thought you'd finished with them when you left the job.'

'Then the same applies to you,' I said. He shrugged his shoulders in agreement. 'I suppose so, there's not much else on retirement when the garden's done, than to come an' have a look round these old crocks.'

His eyes took on a sparkle. 'I'll tell you what, though, it's a young fella's job these days. They was interviewing an Inter-City driver on the box the other night. I'll bet he wasn't more than thirty-five.'

So what was the difference between a diesel and a steam driver? The latter reached middle age before he handled an express regularly.

Jack broke my thoughts. 'Makes you think though, don't it? Those Inter-City trains go a damn sight faster than ever the 'Castles' did, so why all the fuss about being an old man before you handled them.'

Jack was a born cynic who constantly underestimated his station in life, and there were quite a number of enginemen in company with him. 'My boy will never do this for a living,' they would say on a night-shift, and many devoted years of self-denial in steering their sons through to university when that was a rich man's pastime. I gave the old driver a pat on his bent shoulders. 'Come on now, Jack, you'd do it all over again, given a chance.'

His head turned slowly, taking in the full length of the rusted old '28', then he faced me with the slightest trace of a smile. 'Well not on bloody diesels, I wouldn't.'

And that just about sums it up. A new breed of men and locomotives have taken over from the steam men and their machines.

I often meet firemen of my railway days – engine drivers now in their smart suits, who make a rather strained expression when I ask them how they compare the diesel life with the steam life.

'Not like the old days,' say most of them. 'It was always you and the loco. There were good ones, who seemed to take a delight in working with you. There were others who just wanted to break you.'

'Trouble is, we're all the same now,' say others. 'There's no character in the engines, and there's no characters among the chaps that work them!'

But who knows? I never thought that one day I would say proudly 'I was a railway fireman.' Perhaps the day will come when a diesel man will recall the past with similar pride.

Index

Index

Index